ETHICS IN MODELING

Titles of Related Interest

Boyle, D.   Strategic Service Management

Bryson, J.M.   Strategic Planning for Public Service and Non-Profit Organizations

Carlsson, C.   Decision Support Systems

Giarini, O.   The Emerging Service Economy

Thorelli, H.B. and Cavusgil, T.   International Marketing Strategy

Zurawicki, L.   International Countertrade

Related Pergamon Journals – sample copy available on request

Accounting, Management and Information Technology

European Management Journal

International Business Review

Long Range Planning

ETHICS IN MODELING

Edited by

WILLIAM A. WALLACE

Rensselaer Polytechnic Institute, Troy, NY 12180, U.S.A.

PERGAMON

U.K.	Elsevier Science Ltd, The Boulevard, Langford Lane, Kidlington, Oxford, OX5 1GB, U.K.
U.S.A.	Elsevier Science Inc., 660 White Plains Road, Tarrytown, New York 10591-5153, U.S.A.
JAPAN	Elsevier Science Japan, Tsunashima Building Annex, 3-20-12 Yushima, Bunkyo-ku, Tokyo 113, Japan

First edition 1994

Library of Congress Cataloging in Publication Data

Ethics in modeling/ edited by William A. Wallace
p. cm.
Includes bibliographical references and index.
1. Social sciences–Mathematical models–Moral and ethical
aspects–Congresses.
2. Mathematical models–Moral and ethical aspects–
Congresses.
I. Wallace, William A.
H61.25.E87 1994
300'.1'5118–dc20
93-48266

British Library Cataloguing in Publication Data

A catalogue record for this book is available from the British
Library

ISBN 0 08 041930 5

Printed and bound in Great Britain by BPC Wheatons Ltd, Exeter

Contents

Acknowledgements   vii

Chapter 1. Introduction   1
    William A. Wallace

**Chapter 2. How Do the Construction and/or Interpretation
of Models Affect Our Decisions?**   11

Part 1. Uses of Modeling in Science and Society   11
    *John Allison, Abraham Charnes, William W. Cooper
    and Toshiyuki Sueyoshi*

Part 2. An Epistemological View of Decision Aid Technology with
    Emphasis on Expert Systems   37
    Harold D. Carrier and William A. Wallace

Part 3. Models in the Public Sector: Success, Failure and
    Ethical Behavior   58
    John M. Mulvey

Chapter 3. How Do Values Become Incorporated in Models? 75

Part 1. Rhetoric and Rigor in Macroeconomic Models: Populist
    and Orthodox Swings in Latin America   75
    Paul D. McNelis

Part 2. Ethical and Modeling Considerations in Correcting the
    Results of the 1990 Decennial Census   103
    Stephen E. Fienberg

Part 3. The Role of Models in Managerial Decision Making —
        Never Say the Model Says   145
        Vincent P. Barabba

**Chapter 4. What are the Ethical Responsibilities of Model
Builders?**   **161**

Part 1. From Model Building to Risk Management: Evolving
        Standards of Professional Responsibility   161
        N. Phillip Ross and Suzanne Harris

Part 2. On Model Building   167
        John D.C. Little

Part 3. Morality and Models   183
        Richard O. Mason

Part 4. One Sided Practice — Can We Do Better?   195
        Jonathan Rosenhead

Chapter 5. Where Do We Go From Here?   **207**

Part 1. Ethical Concerns and Ethical Answers   207
        Saul I. Gass

Part 2. Responsible Policy Modeling   226
        Warren E. Walker

Part 3. Society's Role in the Ethics of Modeling   242
        Edith H. Leet and William A. Wallace

Appendix A: Authors (Affiliation at Time of Publication)   **247**

Appendix B: Ethics in Modeling Workshop Agenda   **249**

**Appendix C: Ethics in Modeling Workshop Participants
              (Affiliation at Time of Workshop)**   **253**

Author Index   **257**
Subject Index   **263**

ACKNOWLEDGEMENTS

I wish to acknowledge, with great appreciation, the financial support of The Alfred P. Sloan Foundation, and the contribution of our Program Officer, Samuel Goldberg. Sam has lost none of his enthusiasm for modeling from his days as a member of the faculty of the Institute of Basic Mathematics for Application to Business, sponsored by the Ford Foundation at Harvard.

The Department of Decision Sciences and Engineering Systems, in particular, its chair (a modeler himself) Jim Tien, provided additional financial support and encouragement throughout the workshop and the process of preparing the manuscripts for publication. My colleague Tom Willemain acted as a much needed critic for my ideas and reviewed portions of the book concerned with the process of modeling. Edith Leet provided patient expertise in reviewing, proofreading and correcting various versions of manuscripts. Mary Wagner and Pam Bailoni cheerfully took on the task of transferring and reproducing materials, both physically and electronically.

Finally, I wish to thank the scholars who contributed material for the book and participated in the workshop. It is my hope that the book in some way repays them for their time, patience, and willingness to explore a new and often frustrating topic — the Ethics of Modeling.

William A. Wallace
Troy, New York
September 1993

Chapter 1

Introduction

William A. Wallace

In many countries of the world today, mathematical modeling is a major force in decision making for both the public and the private sectors. Policies with wide-ranging and far-reaching effects for millions of people may be based on projections derived from computer analyses of data.

As computer technology advances and the techniques of modeling are more widely taught and understood, more and more people have the facility to build models and to make them increasingly complex. This growth in the numbers, complexity, and significance of models raises many ethical questions such as: What is the proper relationship between the model builder and the model user? Should model builders assume professional responsibility for the results of their models? Do model builders have a responsibility to those affected by the results of their models besides their clients?

These and other questions were raised and considered at the Ethics in Modeling Workshop held at Rensselaer Polytechnic Institute (RPI) on October 12–14, 1989. Sponsored by the Sloan Foundation and RPI's Department of Decision Sciences and Engineering Systems, the workshop brought together professors and practitioners from across the country and Great Britain to present their views. Through presentation of papers and subsequent discussion, they explored many of the ethical dimensions of modeling.

This book contains the papers presented at that workshop, along with other chapters commissioned for this work. In addition, many of the authors have refined and updated the content of their presentations in the intervening period.

THE PROCESS OF MODELING

Modeling is the process of developing and providing an abstraction of reality, i.e. model. The character and result of this process is dependent upon how one intends to use the model. If we wish to use the model to provide us with normative guidelines, i.e. what we ought to be doing, the model will be deductive in nature and may represent a very idealized view of what reality is actually like. Conversely, if we are concerned with describing reality, i.e. developing a model of what is actually occurring, then we utilize various inductive techniques, the most prominent being statistical analysis. In recent years, due to the advent of advanced computer technology, we have been able to develop models that incorporate human judgement and experience. This formalization is typically called a "knowledge representation" scheme. Such representations are in fact models that attempt to describe the thinking processes of humans about a particular problem.

The process of modeling typically goes through the stages of problem conceptualization or formulation, formalization, and exercise and learning. The result is then "used" or implemented.

The formulation phase has traditionally been a murky and understudied aspect of the modeling process.[1] It is here that the modeler works from a "sense" of the problem to assemble the elements and relationships that will comprise the model. Typically, modelers use intuition and experience to make certain simplifying assumptions, and also to address questions such as which factors to include and which to ignore in the model. The formulation stage is crucial because it is both difficult and important to decide whether one formulation is better than another, particularly because the implications of a given form of problem abstraction are difficult to discern so early in the process.

Once formulated in the modeler's mind, the model must be given a concrete form. The formalization selected may take on a variety of forms, such as symbolic, procedural, relational or rule-based.

The next step, after a model has been formulated and formalized, is its exercise in such a way as to yield its benefits. The process is one of interacting with the model to first capture the desired degree of abstraction, and then to address various scenarios based upon the problem or decision situation.

Hazards of Models

Advances in computing, communications and human–machine interface technologies provide the capability to support the process of modeling by either individuals or groups, face-to-face or dispersed. The resulting model can be embedded in "user-friendly" software packages; be integrated into an automatic monitoring and control system for operations; be used to forecast economic impacts for policy analysis; or can be used to determine the credit worthiness of a mortgage applicant.

With the rapid proliferation of information technology and especially easily accessible and user-friendly software, it is inevitable that large numbers of non-experts use sophisticated models with little or no understanding of the process behind them — the assumptions, simplifications and world-view that go into the making of the model. There are three recognizable hazards associated with modeling. Typically, models are designed to handle routine situations. Assumptions of normality in a situation are inherent in most models. In an unusual or exceptional situation, the model provides no support. An attempt to use a model in such a situation by a non-expert could prove catastrophic. A second hazard of modeling is the creation of a model person — or "typing". Today, more and more private information about us goes into files of banks, credit card companies, etc. These companies could use the information to create model people or to "type" people. Since retrieving this information would be fairly simple, this could lead to an atmosphere like a small town, where there is no privacy. A third hazard is the use of a good model for the wrong situation. Called a Type III error, this could lead to immense waste of time, money, and effort, as well as to potentially catastrophic consequences.

OBJECTIVE OF THE BOOK

The objective of this book, and the workshop upon which it was based, is to explore the social and ethical implications of the process of modeling and the use of the resulting models. The agenda of the workshop with the topics and presenters is given in Appendix B, while the participants are listed in Appendix C. Following the workshop, the presenters and participants were asked to comment on the subject itself, addressing four questions (which began Section 2 to Section 5 of the book). Based upon these comments and subsequent discussions among the group, papers were prepared by the presenters and some of the participants.

How Does the Construction and Interpretation of Models Affect Our Decisions?

Allison *et al.*, in the first paper concerned with this question, present the view that "the end of science is publicly available knowledge, i.e. knowledge which is available for all who seek it, and the ethic of science is truth (i.e. fully open honesty) in all parts of any representation intended as a contribution to science. Within science, it is the only ethic. Other ethical issues revolve around uses of science in serving other ends". The thrust of their argument is that there may be a failure to adequately consider that scientific knowledge may depend upon the methodologies used to seek it. Their paper demonstrates this point by reanalyzing data used in the court case involving the breakup of American Telephone and Telegraph Company (AT&T) — with the result that two different methodologies lead to two diametrically opposite recommendations.

They conclude by advising us that non-scientists should be aware that alternative methodologies for modeling used in policy analysis are available and can yield different "scientific" findings!

Carrier and Wallace continue with this point, focusing on the philosophical foundations of statistics, operations research, and expert systems. They develop the idea that a *root metaphor* or a world view exists in both the decision maker and in the decision-aid technologies available to solve a problem. To solve a problem correctly, a decision maker not only must use a decision aid correctly, but first must choose the correct decision aid. When an inappropriate decision aid is used, it creates the possibility of Type III errors: reaching a wrong solution even though the procedures of the technique were followed correctly. As decision aids become both more sophisticated and more user-friendly, they become available for use by more and more naive users, and the possibility of Type III errors increases. In addition, as expert systems become more prominent, it is important to be aware of the philosophical biases of the expert and the knowledge engineer as they interact to develop an expert system. The goal is to increase understanding of the philosophical foundations of decision aids and to increase awareness of the dangers of Type III errors. A framework is given that decision makers can use to help them compare the characteristics of various decision aids and choose the one most appropriate for solving a particular problem.

The interpretation of models is hindered by the fact that many of the techniques used in developing and exercising models are poorly understood by the general public. In his paper, Mulvey notes that this lack of understanding is not unexpected, given the complexity of what he calls "computerized decision procedures". This complexity also contributes to hesitancy on the part of the decision makers to use these procedures — given that it is easier to sell simple solutions.

This situation provides the opportunity for unethical behavior, either by the analyst in an attempt to generate "desirable" or "acceptable" conclusions, or by using the tool as the form of persuasion rather than as a means of arriving at objective results. Mulvey concludes by warning us that unless the modeling profession develops ways of conducting evaluations in an open fashion, "there will be a fatal error which will... precipitate grave consequences and outside pressure for regulation".

How Do Values Become Incorporated in a Model?

McNelis continued the theme of the necessity to analyze the values inherent in a model. He refers to the use of macroeconomic models in providing both guidance and, in many cases, economic resources to Latin American countries. The models used to support the policy analyses that lead to economic presumptions have tended to have either a populist or an orthodox view. Both require ethical critique.

He proposes that macroeconomic modeling, if it is to move beyond the deadlock of populist and orthodox swings, must explicitly build in and analyze variables highlighted by populist models, such as wage and income inequality, while maintaining full recognition of the fundamental fiscal and external constraints in the macroeconomic system.

It must be noted that discussants at the workshop took issue with McNelis's thesis, arguing that the orthodox and populist modelers may not have been intentionally unethical but merely guilty of omitting some variables because they are approaching the problem from different perspectives.

The U.S. Census provides the data upon which political power and federal funds are allocated. It is a source of data for many policy analyses. Feinberg points out that "it is important to understand that no census, no matter how carefully it is carried out or what methodology it uses, can be complete, i.e. without error". He maintains that the problem isn't that the statisticians at the Bureau of the Census are doing a poor job in producing population counts every ten years, but rather that the job they are doing is far more complex than the public understands.

The problem is that the way questions are phrased or variables are defined almost inevitably reflects the political or societal perspective. The result is that any census portrays the values of the society it was designed to benefit. Feinberg supports this supposition with a description of the debates surrounding how the census should be conducted in order to deal with an expected undercount of certain population groups. Although the issue of undercounts has yet to be totally resolved, a statistically sound method was proposed, but a political decision overruled its use.

The necessity to make explicit both the strengths and limitations of a model was reinforced as an ethical issue throughout the workshop as well as in the subsequent paper. Barabba propounded Barabba's Law: "Never Say the Model Says!" What limits model builders from making excessive claims for what their models can do? he asked. What professional restraints exist to curb excessive behavior?

He urges the development of "meaningful and practical ways to enhance ethical behavior", but recognizes that we need diverse views to be considered in model development: model builders do come from exceedingly diverse educational and professional backgrounds. The result is that it would be extremely difficult to formulate and enforce a code of ethics.

His suggestion is that we to some extent let the buyer beware, but have a set of questions that the model user should ask of the model builder. Only through open discussion of the questions can model assumptions and limitations — as well as values — be revealed and assessed.

What are the Ethical Responsibilities of Model Builders?

One of the ethical responsibilities that was agreed upon is that the goal of any model building process is objectivity with clear assumptions, reproducible results, and no advocacy. Ross and Harris report a "story" in which a Congressional Committee wasn't pleased with the results of a consultant's model, and were able to force a revision that produced more agreeable results. They concluded that both parties, the committee and the consultant, shared the blame for this rather flagrant unethical behavior. Political interference must not be permitted to disturb or inhibit the modeling process. If this occurs, the decision makers and their stockholders will not have a good basis on which to make a decision and support its certification and implementation.

The need for close and ongoing communication between the model builder and user was emphasized by Little. His paper reviews the process of modeling and identifies several "pitfalls" that can waylay the model builder. All of these pitfalls can back the model builder into ethical dilemmas. Avoiding these pitfalls requires that there is continuing communication between the builder and the user throughout the modeling process. Not only must the builder provide the user with valid information, but the user must question the builder to ensure that he or she understands the data and assumptions, and how they effect the results. Little also recommends that peer review and open debate take place — on both public issues and in corporate settings.

Mason takes an even stronger position. He describes the relationship between the model builder and user as being based on covenants, three of which are: "(1) to represent reality to her clients adequately, (2) to understand and to incorporate the clients' values into her model in an effective way, and (3) to insure that actions the client takes based on the model have the desired effect".

The difficulty is that covenants cannot be taught in the traditional sense but must be learned through reflection and the assumption of responsibility. Mason goes on to define the "covenant of reality" as one in which the model builder "is entrusted with understanding things as they actually exist in the problem area and representing their most salient features as accurately as possible" and the "covenant of values" as one in which the model builder "is entrusted with the visions, goals, and objectives of her client and pledges to serve these values as loyally as possible".

Dealing with the issue of representing reality is far easier than attempting to translate values into the formalism of a model.

Mason concludes with a warning that advances in information with the proliferation of faster personal computers are increasing the number of model builders. Are we providing means for ensuring that these new model builders will keep the covenants necessary for the successful practice of model building?

The last paper in this section is by Rosenhead, who addresses the issue of whether ethical responsibility should go beyond the relationship with the client and to include the public good. He is concerned that "the modeling profession has acted as if the only organizations in existence were those with hierarchical structures and large resources. Yet there are a plentitude of other organizations, concerned with their members' basic needs (health, education, shelter, employment, environment), which have decision problems no less recalcitrant. Indeed, a case may be made that these community groups have problems which are, if anything, more analytically and practically demanding than conventionally structured organizations. They must scheme and maneuver in a hostile environment with only marginal leverage to protect interests more vital than the next dividend. These community groups are our missing clientele".

Rosenhead urges the profession to work with such community groups, and in so doing, adhere to the ethical principle of equity by ensuring that model building resources are available to all who need them.

Where Do We Go From Here?

The papers in this section had the advantage of being prepared with at least drafts of the papers presented at the workshop. Gass and Walker, with some trepidation but an adventuresome spirit, provide guidelines for professional behavior by modelers. Gass reviews ideas of ethics in other professions and assesses their applicability for the modeling profession. Walker is concerned with policy analysis and the responsibilities inherent in providing recommendations for public policy. He draws upon his experience at RAND in the development of models to support public policy decision making and provides guidelines for responsible policy modeling. Leet and Wallace attempt to summarize the comments of the participants and provide concluding comments.

CONCLUSION

Presentations by the workshop leaders, the subsequent discussion there, and the additional papers in this book make it clear that there are indeed many ethical issues confronting both model builders and the use of models in making public- and private-sector policy decisions. Participants generally agree on some aspects of ethical conduct, such as the need for model builders to be honest, to represent reality as faithfully as possible in their models, to use accurate data, to represent the results of the model as clearly as possible, and to make clear to the model user what the model can do and what its limitations are.

Other ethical issues produce more debate and less agreement:

- Is model building a profession? If so, should the profession set and enforce standards for its members? Does it have the ability to do that?

- Are model builders decision aiders only or do they have a role in making the decisions? Should they take some responsibility for decisions made using the results of their models?

- Do model builders have a responsibility to make their data and procedures public so that others can attempt to replicate their results?

- How should model builders recognize and incorporate the client's values into the model?

- Should model builders incorporate their own values into the model? Can they avoid doing that?

- Do model builders have a responsibility to incorporate the values of others who will be affected by the model besides the client?

- Should model builders work for clients whose values they do not share?

- Do model builders have a responsibility to use their expertise for the good of society by making their services available to groups who might not be able to pay or to pay as much as the traditional clients?

These and similar questions highlight the need for the dialogue about ethics to continue among the professors, practitioners, and users of modeling. Although the workshop provided a setting for such a dialogue, we need to continue and define a vocabulary for exploring ethical concerns.[2] As models become more widely used and the ramifications of their effects become more widely felt, it will become increasingly important for model builders and users to have a clear and strong code of ethics to guide them in making the ethical decisions they will surely have to face.

NOTES AND REFERENCES

1. Willemain, T.R., "Model Formulation: What Experts Think About and When", *Operations Research* (forthcoming).
2. Staudenmaier, J.M., *Personal Communication*, 1990.

Chapter 2

How Do the Construction and/or Interpretation of Models Affect Our Decisions?

Part 1. Uses of Modeling in Science and Society

John Allison, Abraham Charnes, William W. Cooper and Toshiyuki Sueyoshi

INTRODUCTION

The end of science is publicly available knowledge (i.e. knowledge that is available to all who seek it), and the ethic of science is truth (i.e. fully open honesty) in all parts of any representation intended as a contribution to science. Within science it is the *only* ethic. Other ethical issues revolve around the uses of science in serving other ends. Here we adopt John Dewey's definition: an *end* is a collection of *means*. This definition avoids the ambiguity as well as the harm that can result from disassociating the ends from the means used to attain them. It also helps in pointing toward practices and procedures in need of repair. The examples presented are taken from recent legal and legislative failures to consider adequately the ways in which scientific knowledge may depend on the methodologies used. These failures are illustrative of a problem that is becoming increasingly important, because scientific methods are changing rapidly and at different rates in different disciplines. As a start in dealing with these problems, definitions are provided that include data and methodology choices as parts of the sciences and models to be considered, and a Hegelian approach is suggested for uses of science in courts and legislative processes that will help to insure that methodological aspects of the scientific process are adequately considered.

According to "News and Comment",[1] published by the American Association for the Advancement of Science, a juror was quoted as saying, "You can't argue with science" in a case that involved a use of the relatively recently developed technique of "DNA fingerprinting". Actually, science *can* be argued with, or at least it is possible to get scientists to argue with each other. Such an event occurred (outside the courtroom) in this case when, as this article reports,

"...the scientists [involved on both sides of this case] became so concerned over the possibility that the court might be misled by the complexities in science's newest approach to forensic evidence, that they decided to ignore which side they were on and have a mini-science conference. The result of the gathering was a consensus statement that effectively pulled the plug on the scientific evidence that had been offered by the prosecution".

The consensus statement was ruled out by the judge on procedural grounds, because it was "hearsay", and the defense lawyers were forced to adopt other expedients to bring these findings to bear in this case. As one of the defense attorneys notes, "Mostly the evidence has come in without any objection, because the lawyers haven't known how to respond to it".

Part of the problem arises because of the newness of these concepts, and part of the problem arises because of the seemingly arcane scientific methodologies, many of them also relatively new, needed to develop and evaluate the relevant evidence. Another set of problems arises, with accompanying opportunities, when recourse to more than one scientific discipline is needed in order to properly evaluate the evidence that science can supply, and these problems are compounded when the methodologies used in these disciplines are changing.

Substantive findings from the sciences have been incorporated in legal proceedings in more or less acceptable ways for a long time. Witness, for instance, the use of psychologists and psychiatrists to provide expert testimony for opposing parties under the adversary proceedings that characterize Anglo-American law. What about the underlying science methodologies, however, when these are in need of evaluation? An approach to this topic can pose some challenges, especially when scientists do not agree among themselves, and the difficulty is compounded when these methodologies are relatively new even to the sciences that are using them.

Methodology, and in particular the methodology that can be most advantageously used in a particular line of inquiry, is "science business", so to speak, and the differences in methodology that should guide such choices cannot be easily understood or evaluated by persons who are not deeply trained in the pertinent disciplines. Nevertheless, it should be possible to devise procedures and approaches that can make the issues clear for use in law and other areas, or at least a lot clearer than at present, when such methodological choices can yield important differences in the results that flow from using them.

This topic is of increasing importance, we think, partly because ours is an age of increasingly rapid change in the methodologies available for use in the different sciences. Concomitantly, we can expect disagreement and even confusion among scientists as these changes occur. We may, of course, wait for things to settle, but this is not always possible or even desirable, since such delay may have the untoward consequence of failing to make available all of the findings that science can provide.

These difficulties and perplexities are likely to be compounded when methodological changes are occurring, perhaps at different rates in different disciplines, with across-discipline implications that can raise severe questions as to which disciplines should be brought to bear. Note, for instance, that a discipline that might not previously have been pertinent for its bearing on a particular legal proceeding can become pertinent because of the kinds of methodological changes we are considering. It is therefore important to keep such changing possibilities in mind and to suggest ways of bringing them into view as required.

SOME DEFINITIONS

We define a model as follows:

Model: Any system of relations used to represent another system of relations.[2] This definition allows us to comprehend "model equivalences", which can be important when, for instance, a statistician "reparameterizes" an originally formulated model to obtain an equivalent model. It also allows us to identify the methodologies used to obtain solutions or inferences with the set of relations that are used, so that these methodologies also form a part of the model for these purposes.[3]

Issues such as model equivalences (e.g. via statistical reparameterizations) and choices of estimating methods (and even choices of different computer codes) had a bearing on inferences as to whether the evidence showed American Telephone & Telegraph (AT&T) to be a "natural monopoly" in the United States Justice Department-commissioned study by Evans and Heckman.[4] Before discussing that study, however, we draw out some of the further meanings of the above definition as follows:

For one thing, this definition allows us to relate seemingly different model representations to one another. For instance, one uses *geometry* to obtain a pictorial representation of a set of *algebraic* relations. This definition therefore allows us to accommodate transformations from one modeling language to another (e.g. geometry as distinct from algebra), which can then be regarded as equivalent or not, as the case and the intended applications and uses may allow. It also allows us to use different representations in the same modeling language, as in the statistical reparameterizations referred to above.[5]

The relations embodied in a "clay mockup" are also included in this definition so that, for instance, the relation of the steering wheel to the dashboard in a mockup for an automobile can be judged relative to what was specified in a blueprint or a computer printout. The mockup is a "concrete" model, whereas the blueprint (or computer printout) can be regarded as "abstract". The latter, i.e. "abstract models". are the classes we are considering, and these can be further subdivided into models used for decision making and control (sometimes called "normative" models), which are common in operations research and management science, and models used for inference or prediction (sometimes called "descriptive" or "positive" models), which are common in economics and the social sciences.

The issue of what a model is supposed to represent and how it is to be used must also be addressed. We refer to this as an "evaluation" of a model. Thus a model may be evaluated by reference to how well it represents a concrete reality when the model is to be used for prediction, so that it is to be judged by reference to how well it conforms to the part of reality it is supposed to predict. Note, however, that the opposite situation may also occur, as when an already completed floor layout in a factory must be redone because the contractor failed to follow one or more parts of the abstract model embodied in a blueprint that he or she had contracted to follow. In this case, we may think of the model as a "control" to which the reality (e.g. the realized floor layout) must conform.

Bearing these points in mind, we now introduce the following definition:

Evaluation of a model: A statement of the point of view (or standpoint) from which a model is to be judged.[6]

Bear in mind also that the above definitions are intended to incorporate the methodologies from which solutions are to be obtained or from which inferences are to be made from the stated relations. This is important for what follows, because there is a common tendency to divorce the model (e.g. a linear or log–linear model) from the methods used to implement or draw inferences from it.[7]

ACROSS-DISCIPLINE COMPARISON

Tables 1 and 2[8] will help to introduce some of the problems and possibilities of within-discipline and across-discipline approaches to the modeling uses we are considering. The data in these tables provided the basis for estimates derived in a series of statistical studies directed at determining whether AT&T was a "natural monopoly", in which case, its breakup could result in substantial losses of economies that would otherwise be available. Note also that other issues argued in this case,[9] such as "predatory pricing", "cream skimming", etc., and even "barriers to entry", could be handled by regulatory activity, at least in principle.[10] For a natural monopoly, however, no regulatory activity can restore the economies (and concomitant benefits) that will be lost with its breakup. Hence, the natural-monopoly issue was of paramount importance and probably should have received even more attention than was accorded it in *United States* v. *AT&T*,[11] including attention as to whether a suitable model was developed and suitably used with accompanying methodological choices.

Using data compiled in two sources,[12] L. R. Christensen testified[13] that his studies showed evidence of "increasing returns to scale" and thus favored treating AT&T as a "natural monopoly". However, using concepts from "contestable market theory" as [then] recently developed by Drs Baumol, Panzar, and Willig,[14] Drs D. S. Evans and J. J. Heckman restudied this evidence from the standpoint of "economies of scope" (in place of "economies of scale") and found that the evidence did *not* support a contention that AT&T was a natural monopoly.

"Economies of scope" as a multiproduct concept and "economies of scale" as a single-product concept are relatively straightforward issues of economic substance with meanings that are readily interpretable for their bearing in this case. What about issues of methodology? Both Christensen and Evans and Heckman used statistical methodologies and associated models that were of relatively recent vintage and of sufficient sophistication that only a few highly qualified and up-to-date economist-econometricians could be expected to be able to evaluate these methods competently.

Even allowing for this limitation, however, it is rather surprising that nothing more was done by reference to possibilities from other disciplines where rapid evolutions in methodology were also occurring. We refer, in particular, to the disciplines of operations research and management science (OR/MS), where analytical modeling capabilities with associated solution methods had also been evolving.

As part of an experiment to study the possible results of effecting estimates via one of these OR/MS methodologies, it was decided to use a now commonly employed technique known as "goal programming" for these purposes. The idea was to see if this would produce results that might differ from the ones obtained via the econometric methods referred to above.[15] This was accomplished[16] by applying these "goal programming" methods to the same data and the same set of formal relations that had been employed by Professors Evans and Heckman.[17] In other words, only the estimating methods were to be altered, with certain constraints added to ensure conformance to the assumptions made by Professors Evans and Heckman in their study.

In the OR/MS disciplines, goal programming is mostly regarded as a "normative model", or what might better be called a "planning model", in order to describe its uses more accurately.[18] That is, goal programming models are generally formulated on the basis of data that have already been accepted as valid for uses in areas such as multiobjective planning.[19] It had been known for some time, however, that the methods of goal programming could also be used to obtain statistical estimates from already available data.[20] This meant that a way was open to bring to bear the power and efficiency of OR/MS solution methodologies (such as the simplex method of linear programming) in order to deal with the complex simultaneous relations that had to be statistically estimated in this case. Via goal programming and its associated methodologies, the estimation of these simultaneous relations — contrasted with single regression — models could all be accomplished in a direct and relatively straightforward manner, whereas the techniques used by Professors Evans and Heckman made it necessary to proceed indirectly by an approach known as SUR (Seemingly Unrelated Regression) estimates.[21]

Our "goal programming/constrained regression" approach[22] produced rather surprising findings that can be elucidated by reference to Tables 1 and 2. In the footnote to Table 1, the parenthesized values are associated with the erroneous values alongside them that were used by Evans and Heckman in their study. These errors are rather minor,[23] however, compared to the errors that are revealed by effecting a comparison between Tables 1 and 2.

As can be seen, the total costs shown in the first column of Table 1 and the last column of Table 2 are the same. These costs were derived by multiplying the prices shown in the column captioned as P_t by the quantities Q_t to which they apply and then summing the resulting values. The capital, labor, and material prices shown in Table 1 differ from those shown in Table 2. Hence, applying these prices to the quantities shown in Table 2 should have produced a total cost that *differed* from the ones shown in the first column of Table 1. In other words, every one of these observations is wrong, as shown in Table 1, and these are the data on which the Evans and

Heckman estimates were based.

Table 1. Bell System Data Used for Multiproduct Cost Function Estimates

Year	Cost ($10⁶)	Local Output	Total Output	Capital Price	Labor Price	Material Price	R&D Index	Capital Share	Labor Share
1947	2550.68	.41014	(.36642) .34642	.49948	.53566	.66952	.57955	.39552	.49635
1948	2994.94	.45783	(.34642) .37201	.55879	.58236	.75117	.55445	.40430	.48286
1949	3291.06	.48703	.38296	.57440	.60959	.74530	.55261	.41936	.47113
1950	3563.20	.52004	.41592	.61810	.63164	.76525	.56980	.44096	.45352
1951	4047.07	.55560	.46552	.70031	.66926	.81572	.59576	.45338	.44230
1952	4616.23	.59149	.50116	.79500	.70946	.82863	.62057	.46670	.43159
1953	4935.13	.62452	.52271	.80853	.73411	.84389	.63873	.46436	.43614
1954	5258.76	.65669	.55000	.81269	.76134	.85563	.65059	.46596	.42866
1955	5770.47	.70289	.61941	.86056	.80674	.87558	.66162	.47840	.41414
1956	6305.44	.75645	.68394	.88033	.81063	.90493	.68018	.47642	.41045
1957	6351.19	.80355	.74006	.81997	.84824	.93896	.71436	.47138	.41365
1958	6788.40	.84224	.77663	.87304	.85084	.95305	.76830	.50754	.38849
1959	7334.71	.89657	.86274	.91051	.91958	.97417	.83934	.52030	.37321
1960	7912.48	.95314	.93512	.95733	.95979	.99061	.91902	.53120	.36083
1961	8516.46	1.00000	1.00000	1.00000	1.00000	1.00000	1.00000	.54381	.34605
1962	9018.66	1.05411	1.08231	101457	1.03632	1.01995	1.08533	.55077	.33966
1963	9508.12	1.11068	1.17451	1.00832	1.07393	1.03404	1.18984	.55139	.33353
1964	(10524.00) 10542.48	1.15909	1.31715	1.07804	1.12970	1.08451	1.32815	.56240	.32693
1965	11207.00	1.22822	1.47436	1.06139	1.17121	1.10681	1.49998	.55286	.32925
1966	11954.20	1.30609	1.68434	1.04475	1.22827	1.14085	1.16877	.54302	.33698
1967	12710.90	1.38312	1.84266	1.04058	1.29702	1.17371	1.86844	.54079	.34058
1968	13814.10	1.46568	2.05511	1.08325	1.36057	1.21948	2.02744	.54614	.33406
1969	14940.40	1.55869	2.33437	1.04579	1.49416	1.28286	2.16342	.51402	.35802
1970	(16485.80) 16516.87	1.63899	2.53682	1.04891	1.62387	1.35211	2.28416	.49799	.37133
1971	17951.80	1.70956	2.69772	1.04058	1.80415	1.42019	2.40026	.48313	.38304

continued on next page

Table 1. *continued*

1972	20161.20	1.80454	2.96927	1.09157	2.06226	1.47653	2.52124	.47953	.39061
1973	(21221.70)	1.91210	3.31628	1.00312	2.26329	1.56221	2.65447	.44558	.41442
	21190.30								
1974	23168.40	2.00785	3.60503	1.00104	2.51621	1.74061	2.80468	.43407	.42485
1975	27376.70	2.07532	3.86421	1.18939	2.85473	1.91315	2.97195	.46178	.40606
1976	31304.50	2.17307	4.24442	1.32778	3.21920	2.01408	3.15081	.46977	.39508
1977	(36078.00)	2.29155	4.68449	(1.53590)	3.40726	2.12911	3.33422	(.48680)	(.37808).
	34745.33			1.41935				.46712	39259

Note: Parenthesized amounts represent erroneous value for figure immediately to its right.

*Source: A. Charnes, W.W. Cooper, and T. Sueyoshi, "A Goal Programming/Constrained Regression Review of the Bell System Breakup", *Management Science*, 34, 1988, pp. 1-26.

Table 2. Input Quantities and Prices*

Year	Capital		Labor		Materials		Real
t	Q_t	P_t	Q_t	P_t	Q_t	P_t	Cost (10^6)
1947	2101.8	0.480	3065.5	0.413	462.7	0.596	2550.68
1948	2254.9	0.537	3220.8	0.449	528.0	0.640	2994.94
1949	2500.3	0.552	3299.0	0.470	567.5	0.635	3291.06
1950	2645.2	0.594	3318.3	0.487	576.6	0.652	3563.20
1951	2726.4	0.673	3469.1	0.516	607.4	0.695	4047.07
1952	2819.9	0.764	3642.3	0.547	665.0	0.706	4616.23
1953	2949.4	0.777	3802.9	0.566	682.9	0.719	4935.13
1954	3137.5	0.781	3840.3	0.587	760.1	0.729	5258.76
1955	3338.1	0.827	3842.1	0.622	831.2	0.746	5770.47
1956	3550.9	0.846	4141.0	0.625	925.1	0.771	6305.44
1957	3799.3	0.788	4017.1	0.654	912.7	0.800	6351.19
1958	4106.6	0.839	4020.2	0.656	869.1	0.812	6788.40
1959	4361.5	0.875	3861.0	0.709	940.9	0.830	7334.71
1960	4568.6	0.920	3858.2	0.740	1012.2	0.844	7912.48
1961	4819.3	0.961	3822.5	0.771	1100.9	0.852	8516.46
1962	5094.6	0.975	3833.9	0.799	1137.1	0.869	9018.66

continued on next page

Table 2. *continued*

1963	5410.4	0.969	3830.0	0.828	1242.0	.881	9508.12
1964	5713.1	1.036	3950.3	0.871	1280.3	.924	10542.48
1965	6074.5	1.020	4086.3	0.903	1400.9	.943	11206.97
1966	6465.5	1.004	4253.8	0.947	1475.8	.972	11954.19
1967	6874.0	1.000	4329.1	1.000	1507.8	1.000	12710.90
1968	7247.4	1.041	4399.3	1.049	1592.6	1.030	13814.12
1969	7641.6	1.005	4643.3	1.152	1748.9	1.093	14940.44
1970	8144.7	1.008	4889.6	1.252	1896.9	1.152	16516.87
1971	8673.2	1.000	4943.5	1.391	1985.3	1.210	17951.82
1972	9216.3	1.049	4953.0	1.590	2081.1	1.258	20161.19
1973	9809.3	0.964	5035.7	1.745	2214.0	1.331	21190.30
1974	10453.9	0.962	5073.8	1.940	2204.0	1.483	23168.36
1975	11060.5	1.143	5050.7	2.201	2219.6	1.630	27376.69
1976	11525.1	1.276	4983.1	2.482	2465.3	1.716	31304.54
1977	11899.0	1.364	5192.5	2.627	2687.1	1.814	34745.33

Legend: Q = Quantity Index, P = Price Index (using 1967 as price relatives).

*Source: L.R. Christensen, D.C. Christensen and P.E. Schoech, "Total Factor Productivity in the Bell System 1947-49". Christiansen Associates, 810 University Bay Dr., Madison, WI 53705, 1981, pp. 18, 49, and 52.

How this error was made[24] and the arguments that resulted after its discovery are set forth in great detail in the exchange recorded in A. Charnes, W. W. Cooper, and T. Sueyoshi's article, "More on Breaking Up Bell".[25] Here we are concerned rather with the question of how it came to pass that these errors remained unnoticed even though all aspects of the case were supposed to be subject to scrutiny by the considerable talents arrayed on both sides of these adversary proceedings. Our belief is that the failure to uncover these errors occurred because the experts involved (on both sides) were all members of professions or sciences that have grown accustomed to the methodologies of index number construction and statistical regression that are so common in economics and econometrics.

This can be documented in another way if we recount some of what we experienced in the course of completing these goal programming researches, during the course of which the computer printouts repeatedly reported "no solution" as the only possibility. It was this result that made it necessary for us to undertake an examination of the original data from which we have reproduced the pertinent portions in Table 2.[26]

As we were interested in securing a comparison with results from the econometric methodologies used, we modified the model in a way that allowed us to proceed toward this objective and to obtain solutions from our goal programming estimating methods while retaining the closest contact we could arrange with the preceding work by Evans and Heckman.[27] The results obtained from the thus-altered goal programming model were the *reverse* of the results obtained from those reported in the econometric studies by Evans and Heckman in *every year* covered by these studies.[28]

In reflecting on the way the data errors were discovered (along with the reversal of results), we were led to ask how this came about without any of the challenges that would then have been possible in this protracted and vigorously contested trial. Our belief is that confinement to the statistical-regression methods that are commonly used in economics would not have brought forth these discrepancies but would rather have resulted only in inferences with respect to statistical significance, etc., that are customary in these approaches.[29] In the goal programming approach, on the other hand, the discrepancies present in the data led to the "no solution" finding that proved to be the key for discovering these data errors.

Arguments about the validity and deficiencies of these two across-discipline approaches can be found in A. Charnes, W. W. Cooper, and T. Sueyoshi's article, "More on Breaking Up Bell".[30] The point to be made here, however, is that these arguments about the relative advantages and possible consequences of using these different methodologies can (and should) be made explicit and that it would have been better to have the pertinent pros and cons revealed in the course of the AT&T trial, as was not done.

This example from the AT&T hearings is not meant to apply only to economics. It could apply equally well to an exclusive reliance on OR/MS when issues of import are involved, as in the AT&T case and, as should perhaps be explicitly noted, the need for such cross-checking between disciplines is not confined only to the economics/econometrics and OR/MS pairings we have been discussing.

As an example of another possible pairing, we take a similar development from the case of *Texas* v. *New Mexico* concerning Pecos River water, which flows in both of these states and which led to a controversy that was finally settled by the United States Supreme Court in favor of Texas.[31]

Here again, a goal programming/constrained regression approach was matched against state-of-the-art (*circa* 1949) estimating methods used by hydrologists and engineers with resulting differences that we may summarize as follows: In the AT&T case, we showed how the use of a goal-programming/constrained regression approach resulted in the identification of data errors. In the Pecos River case, it resulted in the identification of estimates that, as obtained from their traditional methods, violated basic physical flow relations. In particular, it was discovered that these estimates violated the matchings between inflows and outflows that are required by the "mass balance conditions" of hydrology.

Other differences arise from the uses of goal programming regressions that bear on sensitivity to "outlier" (i.e. aberrant) observations. These differences are well known to statisticians and others, however, and so we leave them aside in order to focus on the inherently unacceptable quality of the estimates that were uncovered from the goal programming applications of the Pecos River data. In fact, it was found that for some of the reaches of this river, the flow would have had to be negative as determined from the more traditional methods, which is to say that water would have had to flow upstream to satisfy the hydrological mass balance requirements.

One may suppose that discrepancies like this could have been discovered without recourse to the use of an alternative method such as these goal-programming/constrained regressions. Studies dating as far back as 1949 had, however, failed to detect these errors,[32] again, we think, because of exclusive reliance on methods that had come to be regarded as [the only ones] having scientific validity. Perhaps this was acceptable in earlier years before the goal-programming/constrained regression approaches had been developed. But in any case, the discovery of unacceptable estimates from the use of these traditional methods formed part of the argument on which the United States Supreme Court ruled in favor of Texas.

We do not pursue further details and ramifications of these studies. Our interest here is rather in how a use of the different methodologies that are common to different disciplines can be made to shed light on issues that might otherwise never be brought into view for consideration in reaching decisions of potentially great import.

This, in turn, raises a question of how to recognize when recourse might be made to such across-discipline methodological approaches. For this, we can make suggestions such as the following.

First, of course, it is necessary to know that such possibilities are present, and second, it is necessary to realize that changes in methodologies do occur over time in the sciences. This necessitates an alertness to the changes in alternative disciplines that might be advantageous to consider. Some degree of familiarity with the nature of such changes and the disciplines in which they occur is desirable, although reliance on advice from expert consultants may be required to some extent. In addition, some notice of such developments can be obtained from the across-discipline references that may occur in different literature. For instance, one might observe the relatively frequent references to economics found in the OR/MS literature and *vice versa*. Finally, some experience and experimentation and the initiation of a tradition of such cross-referencing may be in order for the legal and legislative professions, and the sooner this begins, the better and more easily it will be accommodated. The important point to bear in mind is that methodology as well as substance can make a difference in the outcome of a scientific study in a similar way that procedural as well as substantive law can make a difference in the presentations and the findings of a court.

In the introductory section of this chapter, we referred to the long-established practice in which the adversary procedures of American courts have accommodated substantive differences of findings, such as those of psychiatrists and psychologists, in order to bring them to bear on courtroom trials. What about issues of methodology such as we have been addressing? How have the courts addressed these more difficult-to-assess aspects of science so that they, too, can be adduced and evaluated relative to the judgments that are to be made? Here again we can turn to the dispute between Texas and New Mexico dealing with water flows of the Pecos River. The Special Master (Judge Breitensten) ruled that the issue was purely technical and to be settled by reference to how the estimates were to be made. Relying on customary adversary procedures, however, the judge and the lawyers reduced the choice to either the formula specified in the Pact or to another "later" formula proposed by New Mexico, without recognizing the possibility that *both* might be wrong.

The matter was by chance called to the attention of the Director of Cybernetic Studies at the University of Texas. Representing a different set of disciplines, he provided an alternative approach for the Texas Attorney General's office based on a different methodology from which estimates conforming to the requirements of hydrology were obtained that highlighted the erroneous results. The Special Master had difficulty in understanding this new method and so, on his own initiative, he asked for a simpler method, which he was also unable to understand when it was developed by the University of Texas group and presented to him.

Actually, the procedures outlined under Federal Rules of Evidence allow for a different course, which was brought into play when Judge Meyer succeeded Judge Breitensten as the Special Master. In particular, Judge Meyer immediately appointed a consulting hydrologist (of international repute) to assist him, as is allowed under the Federal Rules of Evidence, and the case was finally settled in favor of Texas.[33]

The law has apparently begun to develop methods of evaluation pointed in directions such as we have been considering. We must not become complacent, however, since (1) Judge Breitensten failed to avail himself of this alternative and (2) it was the intervention of an outsider who brought into view an alternative methodology that highlighted the errors of the previously considered estimating procedures.

WITHIN-DISCIPLINE METHODOLOGICAL DIFFERENCES

It would be wrong to leave an impression that across-discipline differences provide the only possibilities, or that there is no within-discipline checking of alternative methodological possibilities. At least some members of a discipline can generally be regarded as "methodologists" because they are engaged in developing new methodologies, while others are on the lookout for new methodologies that can be imported from other disciplines, possibly in modified form. In addition, methodological changes can alter the boundaries between disciplines and even lead to the development of new sciences, so that possibilities may also need to be considered from this quarter.[34]

An example of such within-discipline methodological explorations along with their possible consequences is provided by reference to a report by R. J. LaLonde.[35] In an article entitled "Evaluating the Econometrics Evaluation of Training Programs with Experimental Data" published in the economic literature, LaLonde brings to bear the techniques of "statistical experimental design"[36] to obtain comparisons with commonly employed econometric methods, which he refers to as "nonexperimental approaches" in order to distinguish them from the alternative of "experimental approaches".

Proceeding via a "field experiment",[37] LaLonde discusses how this approach and its related methodology was used to study a temporary employment training program known as the National Supported Work Demonstration Program. He then proceeds to compare the results obtained via this approach with results secured from econometric methods analogous to those used by Evans and Heckman in their study of AT&T. LaLonde concludes his article by faulting these "non-experimental" (econometric) approaches because "they failed to replicate the experimentally determined results". Thus, in the terminology introduced in the section "Some Definitions", above, LaLonde is effecting his Evaluation from this standpoint.

The Evaluation might also have been conducted from the other side, however, and we did this when we employed a recently developed approach known as "Data Envelopment Analysis" to discover (and disclose) faulty estimates and inferences from a Department of Defense-sponsored field experiment[38] that was used to determine the effects of advertising on recruitment for the military services.

To fit all this into the preceding development, we need only note that evaluations might be undertaken from both standpoints, and that non-scientists should have an opportunity to be made aware that such alternatives are available and can yield different "scientific" findings.

There is an added point to be made from this reference to LaLonde,[39] however, in that these differences also form a part of the scientific validation process in accordance with the following definition:

Science: A systematized body of knowledge along with the methods used to acquire, validate, and systematize that knowledge.

We draw particular attention to the important role we are according to scientific methodology in this definition. Thus, what used to be referred to as "*the* scientific method" is extended here to allow for multiple methodologies. It is these methodologies that characterize the way a scientist holds this knowledge and make it possible to distinguish this activity from that of a mechanic, for example, when both are concerned with the performance of an internal combustion engine. Thus, the scientist may formulate the engine's performance in terms of a set of differential equations because he or she wants to relate this performance to other behaviors (e.g. in thermodynamics), while the mechanic uses gauge readings to determine whether the performance of this particular engine is satisfactory in all respects.

The above definition also makes it possible for us to identify changes in its methodology with changes in a science, including changes in its boundaries and/or the emergence of entirely new sciences and new subdivisions of previously existing services. Hence, the definition draws attention to the need for considering these changes in methodology as well as the need for attending to changes in substance in a world of rapidly changing methodologies.

In general, of course, data differences as well as methodological differences need to be considered, and the two may need to be considered simultaneously. This is illustrated by the following examples.

Some years back, the United States Department of Health, Education, and Welfare (HEW) undertook a series of studies of a Peer Review and Monitoring System known as Professional Standards Review Organizations, which had been legally mandated under an amendment to the Social Security statutes. These organizations, known as PSRO's, represented an effort on the part of Senator Wallace F. Bennett to bring the soaring costs of medical services under control while allowing some flexibility to deal with differences that might be encountered in individual cases. Briefly, the PSRO's provided a review (i.e. an audit) by competent, but independent, medical professionals who would use the results of the review to evaluate the practices of individual hospitals and their staff.[40]

These studies, undertaken at considerable expense by HEW, were subsequently complemented by other studies undertaken by the United States Congressional Budget Office.[41] All of these Federal Government studies used the kinds of econometric-statistical methods we have been discussing and arrived at the same conclusion: No statistically significant effect could be discerned from the use of PSRO's as measured by the length of time that a patient stays in hospitals.

Believing that an alternative methodology might shed additional light on what was occurring, Churchill, Cooper, and Govindarajan undertook an alternative "field experiment" approach.[42] Here we note only the following features: Because the Bennett Amendment and the attendant uses of PSRO's did not apply to Canadian hospitals, the latter were used as a control with further distinctions between: (1) a period before the amendment was passed, and (2) a period of full-scale application thereafter. In addition, distinctions were drawn between patients according to whether they were: (1) over or under the age of 65 years and, (2) had single or multiple ailments to be treated.

None of this was done in the HEW or CBO studies, but in contrast to the total United States coverage in the latter studies, the very limited resources available for this field experiment made it necessary: (1) to confine it to a comparison between a single hospital in the United States and a single hospital in Canada, and (2) to confine it only to the most common surgical procedures used in these two hospitals. Even with these limitations, however, the results of these studies were of potential importance because they reversed every one of the findings of the larger studies. Starting from a common length of stay in both hospitals, the United States hospital began to show a reduced length of stay during the transition period, after which a further reduction was exhibited in the succeeding period of full-scale PSRO operation in the United States hospital. All of the exhibited behaviors thus pointed in the same direction and were statistically significant. The Canadian hospital, on the other hand, displayed an exactly opposite behavior and showed steadily increasing and statistically significant lengths of stay in each of the three periods. Finally, the evidence from this "field experiment", in contrast to the non-experimental HEW and CBO approaches, suggested that the PSRO procedure also had the desired effect of flexibility in its application to individual patients by allowing longer lengths of stay for persons who were over 65 or who had multiple diagnoses.

We think it unfortunate that only the non-experimental approaches were considered when the Bennett amendment was repealed and also think it unfortunate that we could not prevail on any of the Federal agencies to consider any alternative methodology. In summary, we believe that this is the kind of thing that needs to be done not only in legal proceedings but in legislative proceedings as well.

DIFFERENT METHODS — DIFFERENT QUESTIONS

Another topic associated with uses of different methodologies can be brought into view by considering the study by Banker *et al.*,[43] which applied a recently developed methodology known as Data Envelopment Analysis (DEA) to data that had previously been studied with older statistical-econometric methods. The results from the latter study, as reported by Conrad and Straus,[44] resulted in a conclusion that the data did not support the presence of either increasing or decreasing returns to scale in North Carolina hospitals. Only a conclusion of constant returns to scale was found to be consistent with these data. However, the DEA application to these same data yielded results that showed variations in individual hospitals with increasing and decreasing returns to scale sometimes being simultaneously present, for different diagnoses and/or age groups, in the same hospital.[45]

One possible conclusion from these apparently conflicting results is that no scientifically valid conclusion is possible because the results are not invariant to the methods used. In this interpretation, the issue of whether these data show these hospitals to be experiencing constant returns to scale would remain "up for grabs". Other interpretations are possible, however, and can be brought into view by considering some of the estimating principles underlying these two methods. The statistical-econometric methodologies employed by Conrad and Straus[46] proceeded by the principle of "least squares" to obtain an estimating relation that "averages out" the behavior of different hospitals to obtain a single best-fitting relation across *all* hospitals. DEA, on the other hand, optimizes the fit to the observations for *each* hospital so that, in effect, it can be regarded as being directed to the question of which individual hospitals displayed variations in scale returns, and it was also directed to identifying the diagnoses and age groups where these returns were being exhibited.[47]

As this example shows, differences in methodology can yield results that can be interpreted differently, so that help from cognizant scientists may be needed (perhaps from different disciplines) in order to direct attention to these possibilities. Once again, these differences are not restricted to operations research and econometric methodologies, as in the above example. For instance, McKeachie reviewed a long history of experiments in educational psychology directed to studying the effects on learning of varying class sizes, without being able to establish any significant relation between the two.[48] This result was corroborated in another context in the classic study[49] in which the sociologist James Coleman applied standard (single equation) regression–correlation methods to a large body of data on student performances in United States public schools.[50] On the other hand, applying "simultaneous equation (econometric) estimation" methodologies to Coleman's data, Boardman, Davis, and Sanday reported results that showed class size to have a highly significant effect on learning.[51] Referring to the motivation for this study, Boardman, Davis, and Sanday reported that they decided to ascertain whether an application of different methods to these same data might yield different results. As they note, "The failure of [other] researchers to find different results [using Coleman's data] stems primarily from their using similar methodology. In this paper, a new model and a new methodology is applied to the same data to yield new results".[52]

These seemingly contradictory results can be interpreted in a variety of ways by considering the different questions that might have been addressed via the use of these different methodologies. For instance, McKeachie's experimental-method/controlled-laboratory approaches can be interpreted as responding to the following question: Does class size by itself, everything else held constant, affect student learning? The methods used by Boardman *et al.*,[53] on the other hand, can be interpreted as responding to the following rather different question: Can variations in class size be combined with other variables (such as teaching quality and methods, texts, student mixes, etc.) in such a way as to make a difference in learning (i.e. achievement) performances?[54]

Notice that the answer to this second question can be "yes" without necessarily contradicting the results reported in the long series of experiments described by McKeachie.[55] Coleman's model and methodologies were directed in a different way to the same kind of "causal" question (one dependent variable at a time while "controlling" for other variables).[56] Thus, the question addressed by Coleman's methodology can be regarded as being consistent with the results reported by McKeachie, and both of them can differ from the results obtained in the simultaneous relation estimation (to allow for simultaneously operating causes) applied to Coleman's data in the study by Boardman *et al.*[57]

Evidently help may be needed to draw out and clarify such differences. Differences in results obtained by using different methods on the same data need not warrant a conclusion that none of these results are to be relied upon. It is better to be sure that consideration is given to possible differences in interpretation, and this can include differences in the questions being addressed when these can vary according to the methodology used.

Clarification via uses of alternative methodologies can be important for science as well as practice, since it can provide perspective on old questions, as addressed by received methodologies, as well as new questions that can be addressed by a new methodology. This, too, is a way that science can progress as it provides incentives for still further research designed to achieve new methodological advances in response to new problems that thus come into view.

DATA AND DATA AVAILABILITY

The definition of science introduced in the section "Within-Discipline Methodological Differences" is intended to include the data, as well as the theories and the methods, that form a part of the body of knowledge in any science. This has been an acceptable position in the sciences for so long that it may seem superfluous to call attention to this part of a definition of science at this late date. Yet there are disquieting signs that such recognition is necessary and that this necessity is increasingly in need of attention.

The scandals in some of the data collections that supposedly underlie some of the published scientific findings in medicine and psychology are well known and undoubtedly require attention from the scientific community if government regulation of these (as well as other) scientific activities is to be avoided. This is only part of the story, however, since the burgeoning of capabilities associated with developments in electronic computer processing has, in turn, led to a proliferation of data collections with accompanying opportunities for profitable exploration by commercial enterprises.[58] Within limits, this is acceptable and even desirable, provided it does not interfere in any serious way with the requirements for validation and systematization, including validation and systematization by independent investigators, that is included in the above definition.

In the social and managerial sciences, various kinds of access to data may require honoring relations of confidentiality. Many of these requirements can be handled, as has long been done in medical science and engineering, by distinguishing between data that are of interest because of their bearing on scientific generalizations and information that is special to an individual client or company. Thus, a medical scientist describing a newly discovered disease need not identify individual patients when publishing in the scientific literature, and the engineer describing the deficiencies or dangers of a chemical process need not identify the companies where the processes have been troubled since, presumably, this has little bearing on his scientific generalizations. In other cases, the data can be masked or disguised without attenuating the claimed scientific results, and so on. Much will depend on the nature of the claimed scientific results in relation to data, and this may range from results that are only remotely related to the data to results that depend in every way on the data for their validity. In the latter cases, it seems prudent to require a willingness by an author to disclose the data (perhaps with special safeguards) as a condition for publication in the scientific literature. Otherwise, we are in danger of altering a long-standing tradition in which "scientific knowledge is to be regarded as publicly available knowledge and so constantly open to challenge in *all* of its aspects".[59]

To illustrate some of the dangers to this tradition that are currently developing, we may note that one such supplier of information, Institutional Brokers Estimate System (I/B/E/S), believing it was wronged by one piece of research by one user of its information, now requires all researchers to sign a contract under which (a) I/B/E/S will have the right to review and possibly censor the researcher's work and (b) it will also have the right to withhold approval of collaborative research when this may give data access to a colleague.

Something may be said in favor of I/B/E/S, which is, after all, more interested in its commercial advantage than in the advance of scientific knowledge. However, this same sort of thing also appears in the behavior of researchers, albeit in milder form. Consider, for instance, the case in which the American Accounting Association (an association of academic researchers and teachers) adopted a policy bearing on data disclosure that is to be applied to articles submitted to its journals. Under this policy, the authors are required to footnote whether they are willing to disclose their data, but they are not required to disclose it unless the editor deems that the nature of the article requires it.

Even this relatively mild requirement has led to strong expressions of opposition from the Audit Research Committee and the Government and Nonprofit Section of this Association. The argument seems to be that such disclosures (if required) would reduce the competitive advantage possessed by persons who have spent much time and effort in developing the databases on which their claimed results are said to rest. It also seems to be accompanied by a belief that everything that is required for scientific progress is better obtained by requiring others to undertake similar data collections on their own.

Leaving aside the issues of fraud and misrepresentation that were mentioned at the start of this section, one can wonder how the proposed alternatives can assure the absence of errors, even very serious errors, in the resulting collections. As noted in the very important article by Dewald, Thursby, and Anderson,[60] errors of transcription and even errors in computer codes are more common than supposed. We can go further and report our serious concern with the report in that some two-thirds of the investigators who had published articles relatively recently with data-dependent results were unable or unwilling to provide their data when requested to do so by the highly reputable group of scientists concerned with seeing whether they could replicate the investigator's results.[61] To conclude on a perhaps overly mournful note, we quote from page 601 of the article as follows:

"One of the authors of this article was editor of JCMB[62] during most of the period when the articles included in the JCMB *Project* were published. It would be embarrassing to reveal the findings of the *Project* save for our belief that the findings would be little different if articles and authors were selected from other major economics journals. [Indeed] in private correspondence, the editor of another major journal confides that he shares our belief". See also the description of the similar situation in an article entitled "The Missing Crystallography Data" in "News and Comment".[63]

Thus, having begun by noting problems in the use of science by the laity (along with suggestions for dealing with them, we now conclude with the observation that science itself may also be in need of attention and repair.

NOTES AND REFERENCES

1. American Association for the Advancement of Science, "News and Comment", *Science* **245**. Washington, DC: American Association for the Advancement of Science, June 1033, 1989.
2. Cooper, W.W. and Ijiri, Y., *Kohler's Dictionary for Accountants, 6th edn.* Englewood Cliffs, NJ: Prentice-Hall, 332, 1983.
3. See the discussion of what is referred to as the "algorithmic completion of a model" in A. Charnes and W.W. Cooper, "A Strategy for Making Models in Linear Programming", in *Systems Engineering Handbook*, R. Machol *et al.*, eds. NY: McGraw-Hill, 1965, Chapter 26, in recognition of the need to allow for the

appearance of different solution possibilities with different algorithms on what might otherwise be regarded as the same models. See also the discussion in the same chapter of "model equivalences", in general, as distinguished from "equivalences at an optimum", etc.

4. Evans, D.S. and Heckman, J.J., "Test for Subadditivity of the Cost Function with an Application to the Bell System", *American Economic Review*, **74**, 615–623, 1984.

5. This definition also lends itself to the use of "model approximations", which are associated with the techniques described in Chapter 2 of A. Charnes and W.W. Cooper, *Management Models and Industrial Applications of Linear Programming*, Vol 1. NY: John Wiley, 1961, when one model may be used to approximate a possibly different model with perhaps a suitable algorithm needed to ensure a satisfactory approximation.

6. This also includes the criteria to be employed in effecting these judgments, when needed.

7. For further discussion, see A. Charnes and W.W. Cooper, *op. cit.*, Chapter 26.

8. Charnes, A., Cooper, W.W. and Sueyoshi, T., "A Goal Programming/Constrained Regression Review of the Bell System Breakup", *Management Science*, **34**, 1–26, 1988.

9. *United States* v. *AT&T*.

10. For further discussion, see the chapters devoted to these topics in D.S. Evans, ed., *Breaking Up Bell*. Amsterdam: Elsevier Science, 1983. See also Vol. X of *The Journal of Reprints for Antitrust Law and Economics*, 1980.

11. A complete record of this case, including the exhibits as well as the court testimony, may be found in the Telecommunications Library of American University in Washington, D.C.

12. Christensen, L.R., Christensen, D.C. and Schoech, P.E., "Total Factor Productivity in the Bell System 1947–49". Christensen Associates, 810 University Bay Dr., Madison, WI 53705, 1981, and L.R. Christensen, D. Cummings, and P.E. Schoech, "Econometric Estimation of Scale Economies in Telecommunications", in L. Courville, A. de Fontenay, and R. Dobell, eds., *Economic Analysis of Telecommunications: Theory and Applications*. Amsterdam: Elsevier Science, 1981.

13. Christensen, L.R., "Testimony of L.R. Christensen", *United States* v. *AT&T*, C.A. No. 74-169S, 1981.

14. Baumol, W.J., Panzar J.C. and Willig, R.D., *Contestable Markets and the Theory of Industry Structure*. NY: Harcourt, Brace, Jovanovich, Inc., 1982.

15. The original study is reported in T. Sueyoshi, "Goal Programming/Constrained Regression and Alternative Approaches to Statistical Estimation", Ph.D. Thesis, The University of Texas at Austin, Graduate School of Business, 1986 (available

from Ann Arbor, MI: University Microfilms, Inc.), where a detailed discussion may also be found of both goal programming and its uses in this AT&T study.

16. See Charnes, A., Cooper, W.W. and Sueyoshi, T., *op. cit.*

17. See Evans, D.S., *op. cit.*

18. The terms and attitudes that distinguish between normative and descriptive (and other types of models) as employed in the social sciences are less common in the natural sciences, including engineering and medical science, partly because so much of the activity of the latter disciplines is in areas where there is little if any disagreement about the desirability of what is being done. The OR/MS disciplines seem to fall somewhere in between this category and the common social science usages.

19. See Charnes A. and Cooper, W.W., "Goal Programming and Multiple Objective Optimizations", *European Journal of Operational Research*, 1, 39–54, 1977 for a survey of different goal programming model forms and their uses in planning. See also A. Charnes and W.W. Cooper, "Goal Programming and Constrained Regression: A Comment", *Omega*, 4, 403–409, 1975.

20. In fact, this was the form in which it was first developed and used. See A. Charnes, W.W. Cooper, and R.O. Ferguson, "Optimal Estimation of Executive Compensation by Linear Programming". *Management Science*, 1, 138–151, 1955.

21. See Zellner, A., "An Efficient Method of Estimating Seemingly Unrelated Regression Equations and Tests for Aggregation Bias", *Journal of the American Statistical Association*, 348–368, 1968.

22. Charnes, A., Cooper, W.W. and Sueyoshi, T., *op.cit.*

23. And perhaps even to be expected since, as reported by W.G. Dewald, J.G. Thursby, and R.G. Anderson in "Replication in Empirical Economics: The Journal of Money, Credit and Banking Project", *The American Economic Review*, 587–603, 1986, "such errors are more commonplace than rare." The need for providing access to the data used in empirical studies in the interest of science (as well as society) is discussed below.

24. Briefly speaking, these errors were caused by shifting from the 1967 base period used for the prices in Christensen to the 1962 base period used for the prices in Evans and Heckman and then neglecting to adjust the total costs. See the 1961 and 1967 rows in the two tables.

25. Charnes, A., Cooper, W.W. and Sueyoshi, T., "More on Breaking Up Bell", CCS Report 590, Austin, TX: The University of Texas at Austin, Center for Cybernetic Studies, March 1988.

26. We are grateful to Dr. Christensen for making available the data that he and his associates had assembled and processed in an extended study of AT&T records and reports.

27. Evans D.S. and Heckman, J.J., *op. cit.*

28. We also conducted further studies that applied these goal programming/constrained regression studies to correct data with inferences that also differ from those reported in Evans and Heckman. We do not discuss this here, however, because these corrected data differed from the set used by Professors Evans and Heckman in effecting their econometric estimates, and hence this topic is not directly pertinent to the issue we are discussing.

29. To be sure, many of the introductory texts in statistics and econometrics explicitly emphasize the desirability of examining the data, but this is quickly replaced by an emphasis on technical matters such as tests of statistical significance, etc.

30. Charnes, A., Cooper, W.W. and Sueyoshi, T., "More on Breaking Up Bell", *op. cit.*

31. See Senate Document 109, 81st Congress, 1st Session, August 19, 1949, for background detail on the Pecos River Water Compact between Texas and New Mexico.

32. See Duffuaa, S.O., "On Some Network Distributions and Network Problems", Ph.D. Thesis, Austin, TX: The University of Texas at Austin, Graduate School of Business, May 1982. (Available from University Microfilms, Inc., Ann Arbor, MI). These errors were also not discovered in the very detailed review reported in J.R. Erickson, P.P. Hale, J.J. Vandertulp, C.L. Slingerland, and J.C. Anderson, *Report of Review of Basic Data*, submitted to the Engineering Advisory Committee to the Pecos River Compact, 1962. Unpublished Report.

33. Details of the settlement and the need for interdisciplinary cross-checks even at that stage are not covered here.

34. Witness, for example, the current activities in connection with "chaos theory".

35. LaLonde, R.J., "Evaluating the Econometric Evaluation of Training Programs with Experimental Data", *The American Economic Review*, 76, 1986.

36. These methods were originally developed for use in the agricultural experiments conducted at the Rothamstead Agricultural Experiment Station in England and exposited by R.A. Fisher in *The Design of Experiments, 4th edn.* Edinburgh: Oliver and Boyd, 1947.

37. I.e. an experiment conducted in the field rather than in a laboratory setting.

38. See Brockett, P.L., Charnes, A., Cooper W.W. and Golany, B., " A Critique of the Wharton Center for Applied Research Study of the Effects of Advertising on Recruitment", *CCS Research Report* 546, Austin, TX: The University of Texas at Austin, Center for Cybernetic Studies, July 1986, and A. Charnes, W.W. Cooper, B. Golany, D. Thomas, and R. Halek, "A Two-Phase Procedure for Efficiency Analysis and Evaluation of Advertising in Army Recruitment Activities", *Research Report CCS* 521, Austin, TX: The University of Texas at Austin, Center for Cybernetic Studies, November 1985. A subsequent study by

the RAND Corporation, which also used nonexperimental (econometric-statistical) methods reached similar conclusions. See J.N. Dertouzos, "The Effectiveness of Service and Joint Advertising: A DOD Perspective", *WD-4249-Four Management Personnel*, Santa Monica, CA: The RAND Corporation, January 1989.

39. LaLonde, R.J., *op. cit.*

40. See Churchill, N.C., Cooper, W.W. and Govindarajan, V., "Effects of Audits on the Behavior of Medical Professionals Under the Bennett Amendment," *Auditing: A Journal of Practice & Theory*, 1, 1982, for a more detailed description that fits the PSRO process into a more general audit process.

41. *Ibid.*

42. *Ibid.*

43. Banker, R.D., Conrad, R.F. and Straus, R.P., "A Comparative Application of DEA and Translog Methods: An Illustrative Study of Hospital Production", *Management Science*, 30–44, 1986.

44. Conrad R.F. and Straus, R.P., "A Multiple-Output Multiple-Input Model of the Hospital Industry in North Carolina", *Applied Economics*, 341–352, 1983.

45. Banker, R.D., *op. cit.*, 43.

46. Conrad, R.F. and Straus, R.P., *op. cit.*

47. See Banker, R., Charnes, A., Cooper, W.W., Swarts, J. and Thomas, D., "An Introduction to Data Envelopment Analysis and Some of Its Uses", *Research in Governmental and Nonprofit Accounting*, 5, 187–210, 1989, for further discussion of DEA-regression comparisons and possible complementary uses.

48. McKeachie, W.J., "Motivation, Teaching and Learning", in M. R. Jones, ed., *Nebraska Symposium on Motivation and Learning.* Lincoln, NE: University of Nebraska Press, 1965.

49. Coleman, J.S., Campbell, E.Q., Hobson, C.J., McPartland, J., Mood, A.M., Weinfeld, F.D. and York, R.L., *Equality of Educational Opportunity*, Washington, DC: U.S. Department of Health, Education and Welfare, U.S. Government Printing Office, 1966.

50. This corroboration needs to be interpreted broadly, however, since Coleman *et al.*, refer to achievement, e.g. as measured by Metropolitan Achievement Tests and similar scores developed from tests devised by the Educational Testing Service.

51. Boardman, A.E., Davis, O.A. and Sanday, P.R., "A Simultaneous Equation Model of the Educational Process", *Journal of Public Economics*, 7, 23–49, 1977.

52. *Ibid.*, 24.

53. Boardman, A.E., Davis, O.A. and Sanday, P.R., *op. cit.*

54. See Boardman, A.E., "A Simultaneous Equation Model of the Educational

Process", Ph.D. Thesis, Pittsburgh, PA: Carnegie Mellon University, School of Urban & Public Affairs, 115, bottom, 1975. Also available from University Microfilms, Inc., Ann Arbor, MI.

55. McKeachie, W.J., *op. cit.*

56. See Charnes, A., Cooper, W.W., Learner, D.B. and Phillips, F.Y., "Management Sciences and Marketing Management", *Journal of Marketing*, **49**, 1985, pp. 93–105, for further discussion of these differences and their importance for pure and applied science.

57. Boardman, A.E., Davis, O.A. and Sanday, P.R., *op. cit.*

58. Clarke D. and Moult, W.H., *Report 583-048: Marketing Research Services: 100 Major Commercial Suppliers*. Boston: Harvard Business School, 1982, list more than 100 firms supplying marketing information in the period 1980–81. See also A. Charnes, W.W. Cooper, D.B. Learner, and F.Y. Phillips, *op. cit.*

59. Quoted from a letter by W.W. Cooper to the Chair of the Government and Nonprofit Section of the American Accounting Association, who petitioned for the withdrawal of even a very weak requirement that authors disclose whether they are willing to make data available to others and, if so, to say how the data might be accessed.

60. Dewald, W.G., Thursby, J.G. and Anderson, R.G., *op. cit.*

61. *Ibid.*, see Table 1 on 591.

62. JCMB = *Journal of Money Credit and Banking*. This was the journal that provided the focus of this project which, with NSF support, attempted to replicate the results of articles published in its pages without much success, even in cases where the authors had complied with the request for their data.

63. American Association for the Advancement of Science, "News and Comment", *Science*, **245**. Washington, DC: American Association for the Advancement of Science, 1179–1181, September 1989. See also the section of Proprietary Software in P.T. Boggs, H.F. Jackson (Chair), S.G. Nash, and S. Powell, "Report of the *Ad Hoc* Committee To Revise the Guidelines for Reporting Computational Experiments in Mathematical Programming", which presents special problems because of issues of possible loss of commercial advantage occasioned by the "full disclosure" requirements for publication in the scientific literature. This was also a problem addressed in the original ACM Guidelines. See *ACM Transactions on Math. Software*, **5**, 193–203, 1979, for the original guidelines, where a distinction was drawn for the criteria to be used with respect to "full disclosure" in the scientific as distinct from the consulting–commercial literature.

64. Coleman, J.S., "Equal Schools or Equal Students", *The Public Interest*, **4**, 70–75, 1966.

65. Conrad, R.F. and Straus, R.P., "A Multiple-Output Multiple-Input Model of the Hospital Industry in North Carolina", *Applied Economics*, 341–352, 1983.

66. Carroll, V.P., "DOD Advertising Mix Test: Comparisons of Joint-Service with Service-Specific Strategies and Levels of Funding", *A Report to the Assistant Secretary of Defense for Force Management and Personnel*, Philadelphia: The Wharton Center for Applied Research, July 1986.

67. Charnes, A., Cooper, W.W. and Golany, B., "Relative Effects of Service Specific and Joint National Advertising in Marine Corps Recruitment Activities", CCS Report, Austin, TX: The University of Texas at Austin, Center for Cybernetic Studies, June 1986.

68. Charnes, A., Cooper, W.W. and Golany, B., "Relative Effects by Data Envelopment Analysis of Service Specific and Joint National Advertising in Navy Recruitment Analysis", CCS Research Report, Austin, TX: The University of Texas at Austin, Center for Cybernetic Studies, July 1986.

69. Charnes, A., Cooper, W.W., Golany, B., Halek, R., Klopp, G., Schmitz, E. and Thomas, D., "Data Envelopment Analysis Approaches to Policy Evaluation and Management of Army Recruiting Activities: Tradeoffs Between Joint Services and Army Advertising", CCS Report 532, Austin, TX: The University of Texas at Austin, Center for Cybernetic Studies, March 1986.

70. Charnes, A., Cooper, W.W., Golany, B., Seiford, L. and Stutz, J., "Foundations of Data Envelopment Analyses for Pareto-Koopmans Efficient Empirical Production Functions", *Journal of Econometrics*, **30**, 91–107, 1985.

71. Charnes, A., Cooper, W.W. and Sueyoshi, T., "Least Squares/Ridge Regressions and Goal Programming/Constrained Regression Alternatives", *European Journal of Operational Research*, **27**, 146–157, 1986.

72. Engineering Advisory Committee to the Pecos River Compact, *Senate Document 109*, 81st Congress, 1st Session, August 19, 1949.

Chapter 2

Part 2. An Epistemological View of Decision Aid Technology with Emphasis on Expert Systems*

Harold D. Carrier and William A. Wallace

INTRODUCTION

Advances in information technology have resulted in decision support software that provides modeling capabilities using statistics, operations research, and, more recently, artificial intelligence (AI) in the context of expert systems. Many of these packages are designed to be user friendly and to minimize the amount of specialized knowledge required for their use. In so doing, they provide little guidance for the user in determining the appropriateness of the supporting philosophical theory or mathematical model for the decision situation. While ease of use is often the predominant criterion in the selection of a software package, its philosophical point of view may make the package unsuitable for the problem or decision situation.

As noted by Mulvey and Blount-White,[1] even a professional modeler needs to ask questions concerning the biases and assumptions inherent in the choice of a method and to compare alternative models. We hope to cast some light on the issue by exploring the philosophical foundations of models derived from statistics, operations research, and expert systems.

We start by exploring the philosophical similarities and differences of these decision aid technologies and show that each is based upon inherently different epistemological foundations. Churchman[2] has written extensively about the philosophical underpinnings of operations research and statistics,

* This paper is a slightly revised version of a paper by the same name published in *IEEE Transactions on Systems, Man and Cybernetics*, Vol. 19, No. 5, September/October 1989, pp. 1021–1029.

but little work has been done on the philosophical foundations of AI, especially expert systems. A number of articles focus on the ethical problems inherent in AI,[3] still others on the cognitive psychological issues from an epistemological perspective,[4] yet ours is the first philosophical analysis that compares and contrasts expert systems with other decision aid technologies.

We focus throughout on the problem of providing appropriate decision support to managers or their agents. The models employed in this task are assumed to be computer-based and not "competent" to make the decision, i.e. the decision cannot be automated. This approach uses Ginzberg and Stohr's definition of a decision support system (DSS) as "... a computer-based information system used by decision makers to support their decisionmaking activities in situations where it is not possible or not desirable to have an automated system perform the entire decision process".[5]

We further assume that the decision aid must contain a facility for modeling and constructing abstractions of reality. Note that an electronic spreadsheet provides a structure, albeit simplistic, that is itself a model and provides modeling capabilities.[6]

Statistics, operations research, and expert systems represent three different views of the reality of a decision situation: descriptive, prescriptive, and rescriptive. Statistics can be characterized as descriptive: an attempt to describe and define parameters, to explain portions of variance, and to account for differences between and within samples.[7] Operations research is prescriptive: it attempts to define normative behavior, or what Churchman[8] refers to as "what ought to be". Expert systems are rescriptive: they attempt to rewrite or "rescribe" the decision methods intuitively held by experts and authorities.

Some researchers have proposed integrative frameworks for decision support systems.[9] However, none of their research addressed the philosophical foundations of the decision aid technologies. In the absence of these foundations, the limitations and partial nature of models tend to be overlooked and the models regarded as normative in and of themselves. Most professional modelers do understand the mathematical basis and, at least implicitly, the philosophic implications of their decision models. For an example, let us consider a warehouse location problem. An operations research professional may develop an integer programming model, while his or her statistical counterpart may use an econometric analysis. One knowledgeable in the technology of expert systems may use a knowledge engineering shell to analyze the problem. None of these approaches is wrong as long as the researcher comprehends the assumptions inherent in their use.

With the proliferation and availability of software, however, there are more and more naive modelers who have little or no appreciation of the underlying assumptions, both philosophic and mathematical, of the decision aids they employ. This problem has been poignantly expressed by Winner, who quotes an employee as saying, "I don't comprehend it, but I'm sure the system does".[10] This leads to problems of validity, especially Type III errors, i.e. arriving at a wrong solution because an inappropriate empirical method was selected for the problem.[11]

Raiffa,[12] citing Tukey, suggests that part of the cause of Type III errors is the gap that exists between the decision maker and the analyst. This gap can be perceived as one inherent in the communications process with a limited amount of feedback and problem definition. In the problem definition stage, he suggests that communication between the decision maker and the knowledge engineer is critical, "to gain some reasonable perspective and sensitivity for a problem area".[13] This is important, he concludes, to ensure that the analysis addresses the real problem.

Davis and Olson[14] cite an example of this type of error in the use of a forecasting technique for a company that sold earth-moving machinery. The company used regression analysis and found that their past sales showed a constant percentage of growth. They extrapolated these data forward and decided to build a new plant to meet the projected demand. The plant closed after sales were far behind those projected.

The company carried out the regression analysis correctly. Their error was in choosing regression analysis as a forecasting technique. They did not understand that the sales of their equipment are not associated with time but rather with demand for construction and the need to replace existing equipment. Because the regression analysis was based on time, the company received a bad forecast and consequently made a decision that proved to be very costly to them.

There are a variety of reasons for errors of the third kind that are due to poor methodological decisions. In addition to selecting an inappropriate decision aid, expediency can cause a misdiagnosis or poor specification of problem. At the other extreme, logical inadequacies at the strategic level can cause misdirection in problem solving.[15]

As early as 1954, Simon[16] urged caution in the use of decision aids; Morris[17] focused on the difference between the "teaching of models" and the "teaching of modeling", citing the underlying differences between the context of discovery versus the context of justification; more recently, Mulvey[18] has attempted to develop criteria for modeling that clearly assert that care must be used in employing decision aids for policy and analysis.

To summarize, decision aid technology is involved in the process of knowing reality. Its purpose is to gather and interpret data and to present some facet or picture of "reality" for the user of the information. Just as telescopes are designed to extend the sensory capacity of humans, decision aids are designed to extend their cognitive capacity. This means that implicit in all decision aid technologies is a theory of what knowing is and hence what reality is. An epistemology is implicit in all decision aid systems. Also, it means that if a decision aid is to extend human cognitive capacities accurately, the assumptions about how we perceive reality and what reality is must correspond to the actual experience of human knowing. One of the tasks of a research agenda should be to render explicit the cognitive assumptions and correspondence of current decision-aid technology. It is our contention that knowledge of the philosophical underpinning of a decision aid can provide a context for selecting appropriate tools: matching the tool to the problem. This can be thought of as a meta-decision process similar to specifying procedures and rules before the modeling process is undertaken.[19]

EPISTEMOLOGY AS A METHOD OF INQUIRY

Often we seem to know what reality is without having the ability to express it in words. We seem to have knowledge about what truth or courage are, for example, yet cannot give them adequate definition. These fuzzy areas of knowledge occur often because much of what we know is intuitive and has not yet been reflected in our consciousness.[20] Philosophical inquiry begins from this intuitive knowledge and is the process of consciously reflecting upon what we know intuitively to make it explicit and demonstrable.

For Descartes[21] this reflection started with doubt, confronting his own uncertainty. In a sense, Descartes placed brackets around what must be reflected and attempted to see a thing as it really is. Since what we know is rooted deeply in the human perceptual process and known directly through consciousness,[22] the notions of world view and intentionality become the primary tools for philosophical investigation.

Epistemology is the study of knowledge and the justification of belief.[23] Epistemology, as a method of inquiry in the decision sciences, has been used by Helmer and Rescher,[24] Churchman,[25] Mason and Mitroff,[26] Mitroff and Sagasti,[27] and Mitroff *et al.*[28] Helmer and Rescher[29] stress the importance of epistemological research in the management sciences because of the inexact nature of the human decision making process. Due to their inexact nature, the management sciences have not been able to create and verify a theory of decision making. On this basis they suggest that, methodologically, epistemology is an important factor in establishing the management sciences on a par with the physical sciences.

METAPHOR AS A TOOL

Two qualities distinguish human activity from other forms of life: (1) the ability to devise and understand symbolic representation, and (2) intentional behavior. Human symbolic activity has received a great deal of attention in the philosophic and psychological literature, from Carnap's logical syntax to Goethe's "ur-symbols" of culture to Freud and Jung's psychoanalytic interpretations. An important result for decision aids is the productivity of symbols, i.e. "... they yield more dividends ... than the conceptual capital originally invested ... furthermore, symbolic universes follow autonomous laws ... as it were, a logic or life of their own, transcending the ... human individuals who created them".[30]

The use of symbols is linked to the concept of intentionality, which deals with human consciousness. Intentionality, as used here, should not be confused with the notion of "doing something on purpose", as in, "he intentionally slammed the door in the salesperson's face". Intentionality is the relationship between consciousness and the reality to be perceived.

According to Descarte's *cogito*, we are aware of ourselves, and as expanded by Husserl,[31] we are aware that there is a physical reality outside ourselves and that being human is being open to the world. The concept of intentionality unites the world of ideas, which has historically created the dualism of mind and body, with that of the world of science, which has created the notion of objective versus subjective reality. This occurs through the perceptual process; the notion of intentionality leads us to the idea of viewpoint and perspective as epistemological *a priori*.

The essence of metaphor, in the broadest sense, is seeing something from the viewpoint of something else; it is an attempt to integrate diverse phenomena without destroying their unique differences. Metaphor transfers a term from one system or dimensional level of meaning to another; it is a *gestalt*.[32] Metaphor demonstrates the intentional nature of the perceptual process; it is the conscious application of relationships between the knower and the known. A computer program that plays chess or helps a physician make a diagnosis, for example, is based on a mathematical metaphor of human cognitive processes. Metaphor is a connection between our consciousness and some tangible or sensible phenomenon, whether it be in a poem or in the heuristics of a diagnostic procedure. Metaphor is therefore an important concept in understanding the nature of the decision sciences and their technologies.

Pepper[33] distinguishes four basic philosophic stances or "root metaphors" that have systematically appeared in Western thought. He contends that all scientific hypotheses are based on one of these root metaphors. Figure 1 shows these major root metaphors in relation to one another. The framework is distinguished along two axes, the analytic–synthetic and the dispersive–integrative.

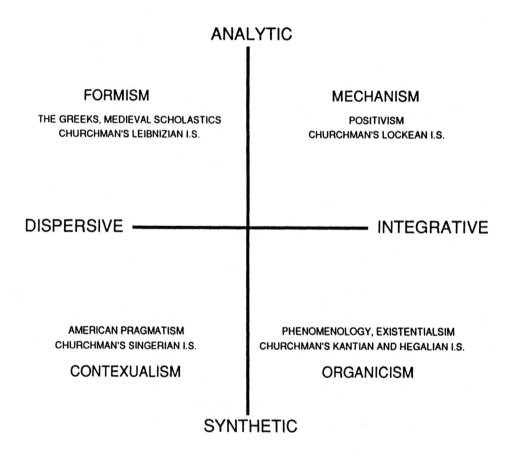

Fig. 1. Pepper's typology of root metaphors.

The analytic–synthetic axis is a continuum of logic employed in seeing the world. Analytic logic uses a step-by-step approach in logical analysis and is dominated by reason. Synthetic logic, on the other hand, attempts to combine separate elements and is dominated by the use of intuition. The dispersive–integrative axis describes the method of inquiry. A dispersive approach defines categories, subdividing items into their smallest units. An integrative approach views items in a systematic way by relating to their relationships and focusing on "the whole picture". By way of analogy, the microbiologist attempts to break down an organism into smaller and smaller components, whereas the biological ecologist views an organism in the framework of a broader biological system.

Pepper believes that the framework serves as a coherent understructure that demonstrates the philosophic foundation of scientific inquiry. His typology includes formism or idealism, whose proponents, including the Greeks and the medieval scholastics, posit a dichotomy of matter versus form or essence; mechanism rooted in British empiricism and currently the foundation of positivism; contextualism, which Pepper claims as the invention of the American pragmatists; and organicism, or an integrative, holistic approach of existentialism and phenomenology.

The implication of Pepper's typology is that the philosophic traditions inherent in science serve as *a priori* filters for analysis and modeling. The decision sciences therefore have a foundation in one or more of these philosophic traditions.

Root metaphors are used as analogies, according to Pepper. Individuals look for analogies or clues from their own experience to make the larger reality comprehensible. Pepper cites the example of Thales, who believed that water was the principle element. Thale's simultaneous dissatisfaction with the beliefs of mythology and the fact that he was a resident of a seaport town totally dependent on the sea for its survival made water his logical choice for the world's most important substance. Individuals are products of their experience, culture, and education. Concepts that have lost touch with their root metaphor, Pepper suggests, become just empty abstractions; they have lost touch with the foundations that made sense out of them.

Pepper's view of root metaphor seems close conceptually to Churchman's[34] typology of inquiry systems and is represented within the scheme of Fig. 1, with Leibniz associated with formism, Locke with mechanism, Kant and Hegel with organicism, and Singer rooted in the pragmatics of contextualism. Churchman's Leibnizian inquiry system is characterized by analytical thought, and the use of categorization or the dispersive method to discover truth. His Lockean inquiry system, although analytical, is juxtaposed from the Leibnizian approach because it is an integrative view of scientific inquiry. The Kantian and Hegalian inquiry systems employ synthetic logic, or what they call the "dialectic" approach, and attempt to integrate categories rather than disperse them. Finally, the Singerian inquiry system, based on the work of the American pragmatist Edgar Singer, is dispersial, yet simultaneously employs the synthetic approach by emphasizing the nature of systems.

Based on Churchman's work,[35] Mitroff and Kilmann[36] have studied the psychological bias of scientists as well as developing a framework for classifying scientific investigation. Additionally, Dunn[37] has suggested that the management science literature has often used constructs that attempt to provide conceptual underpinnings or frameworks. These appear under various names including systems of interpretation, cognitive maps, schema, frames of reference, and construction systems.

MANAGEMENT SCIENCE: THE LOGIC AND METAPHORS OF STATISTICS AND OPERATIONS RESEARCH

Much of what we call management science has a basis in the mechanistic and formistic root metaphors, one having a pronounced bias toward empiricism and logical positivism, the other idealism. The logic of scientific discovery, rooted in these paradigms, uses formal schemes of induction and deduction as paths of inquiry.

A review of the foundational literature of statistics reveals essential agreement with the inductive basis of inferential statistics. Induction enables a researcher to extend experience. As an epistemological tool, it is able to see events and broaden their horizon. The conclusions of inductive arguments have more information than their premises provide.[38] Box[39] and Leonard[40] are in agreement but further differentiate statistical theory into the sampling (or frequentist) school of Fisher, Neyman, and Pearson and the inductive inference of Bayesian statistical theory.[41] Barnett[42] distinguishes statistical induction from probability theory, which he concludes has its logical foundation in deductive logic:

> Any statistical analysis of ... data rests on the construction of an appropriate model and on the assumption of its adequacy ... the model embodies a great deal of structure, being unspecified only to the extent of the unknown values of a few parameters. Again the real situation has been replaced by the model, and the ideas of probability theory may be applied to deduce the characteristic by assumption from the real situation. The use of a statistical procedure, e.g. analysis

of variance, again attempts to reverse this process, by using the observed data to reflect back on the model...[43]

Operations research, like probability theory, uses deductive logic as a basis of its decision making technology. Deductive logic as a tool of epistemology has as its goal justification. The relationship between deduction and justification is such that, if the premises are true, then the conclusion must be true, and being true, they are justified.[44] Churchman[45] suggests that scientists applying this decision technology try to come up with specific facts and make general recommendations. This can be done by examining the past behavior of decision makers and attempting to arrive at some rational rules of general consistency. Relying on the possibilities of the subjunctive mood, one can attempt to establish normative behaviors of what "ought to be". Operations research in the Churchmanian typology of inquiry systems[46] is essentially Leibnitzian, i.e. it is centrally focused on information derived from models and attempts to propose "single best models" from formal symbolic systems. The Leibnitzian inquiry system uses clearly defined and well-structured models to generate the collection of evidence.

TOWARDS A PHILOSOPHIC FOUNDATION FOR EXPERT SYSTEMS: AN EPISTEMOLOGICAL PERSPECTIVE ON EXPERT SYSTEMS

Through the Renaissance and the Age of Enlightenment that followed, the clock became a universal metaphor. Leonardo's sketch books and the clockwork precision of the theories of Copernicus, Kepler, and Newton held the popular imagination. This age of mechanism prefigured the rise of the Industrial Revolution, in which mechanics changed the reality of human existence. It was an age of technological growth; a mere 300 years earlier, the great Gothic cathedrals were built with only the sophistication of the measuring rod, square, and plumb. Mechanism has been the dominant metaphor for nearly 500 years and is used to explain everything from human digestion to plate tectonics.[47]

It is our assertion that a paradigm shift in science[48] has occurred and that the prevalent metaphor is now recursive,[49] information-oriented,[50] holistic,[51] and systemic.[52] In other words, the computer has replaced the clock as a universal metaphor. Turing[53] was perhaps the first to realize the implications of this paradigm shift when he suggested that it was possible for machines to "think".

The computer's sense of logic, however, although rational and constructed by human beings, is itself sufficiently abstract as to require a metaphor for comprehensibility. Can a comparison be made between the computer and the brain? Sutherland,[54] in his critique of AI performance, suggests that it is dangerous to assume a correlation between the neurocomplexity of the human brain and even the most sophisticated computer. This is an inherent epistemological problem, for computers with their relatively linear logic cannot approximate the mind, which is a simultaneous multidimensional gestalt[55] and whose logic is not only non-linear but perhaps relativistic and non-Euclidean as well.[56] The development of AI leads to some inherent philosophical issues. What is the difference between human cognition and decision making? What are the problems inherent in attempts to replicate it with machine logic? The principal problem is the relationship between the knower and the known. This relationship is intensified by the complexity of the variables involved: on the one hand, the human brain with billions of interconnecting neurological components and on the other, machines that can do billions of arithmetic computations in a second. Perhaps the most convincing argument in this vein comes from Gödel's incompleteness theorem,[57] suggesting that the computer is limited by the inherent finitude of mathematical logic itself. The limits of computation are just as real as the limits of the cognitive capacity of the brain.

Expert systems, as opposed to other AI technologies, have a strong bond to the human decision maker. The general characteristics of an expert system include a high level of expertise, the ability to do predictive modeling, and an institutional memory, i.e. the system incorporates the policies of an organization that have been developed by the interaction of key personnel. Methodological problems that exist in developing an expert system include defining the correct problem space, choosing the appropriate expert, using the proper tools for knowledge engineering, developing an appropriate representation, and managing the interaction of these factors.[58]

Waterman[59] addresses the critical interaction of the knowledge engineer and the human expert as the knowledge engineer attempts to capture the expert's procedures, strategies, and rules of thumb and use that information to build an expert system. One critical aspect is how one defines the rather subjective term "expert". Johnson[60] suggests that an expert is set apart by training and experience that enable him or her to cut through much of what is irrelevant, as well as by knowledge of the "tricks of the trade".

There are various epistemological theories that could be applied to the foundations of expert systems, such as inductive versus deductive or positivistic versus phenomenological. We believe, however, that often expert systems arise pragmatically and sometimes unsystematically from the unique perspective of the individuals involved in the creation of the knowledge-based system. This in itself poses philosophical difficulties, for knowledge engineers may have given little thought to their unique world view and may unwittingly cause Type III errors.

Perhaps the most critical aspect, from a philosophical perspective, is the role of the domain expert employed in the development of the system. There is a strong bias towards engineering approaches to problem solving in expert systems, and this is not surprising, given its origins and current applications. However, the problem that arises is equivalent to that posed by Helmer and Rescher[61] concerning the differentiation between the physical and management sciences. Expert systems attempt to mimic the inexact decision processes of a knowledgeable person. The question is whether engineering approaches based upon the physical sciences are appropriate to capture this knowledge.

Mitroff and Kilmann[62] assert that an individual scientist sees the world through a particular psychological perspective that creates bias. They posit four classifications of scientists: the analytic scientist, the conceptual theorist, the conceptual humanist, and the particular humanist. Their thesis explores the external and internal properties of these classifications with an emphasis on the differences employed in logic, the modes of inquiry, the status of science and its ultimate aims, and the personal values of the scientists themselves. We contend that these psychological attributes in domain experts and their influence on the development of expert systems have not been given proper consideration in expert systems research.

The classic studies of Janis and Mann,[63] Morris,[64] and Festinger[65] demonstrate not only the difficulty in understanding the human decision process but also the critical psychological limitations and interactions that qualify the effectiveness of a decision. Even experts find it difficult to articulate what they know. Take, for example, a father teaching his daughter how to ride a bicycle. Riding a bicycle is ingrained within him; he has ridden for 30 years and has not given it much thought. The challenge for him is to give verbal information to the youngster along with the balance provided by a firm grip on the bicycle while she attempts her first awkward pedaling. Perhaps he, as a physicist, could explain the Newtonian processes at work to her, but they would not necessarily provide the information she needs. He could encourage her and give her positive reinforcement about her particular level of performance, but that might not be the kind of technical information she needs in her tentative struggle to keep her balance. Much of what we know is tacitly known and is, as we suggested earlier, prereflected.

Polanyi[66] has carried out an impressive investigation into the nature of tacit knowledge. Citing the work of Gestalt psychologists, Polanyi[67] suggests that we are able to integrate awareness of reality, yet we may not be able simultaneously to identify the particulars that make up this reality. It is perhaps the differentiation between the German *wissen* and *können*, "knowing what" and "knowing how":

> These two aspects of knowing have similar structure and neither is ever present without the other. This is particularly clear in the art of diagnosing, which intimately combines skillful testing with expert observation.[68]

Collins *et al.*[69] have applied the notion of tacit knowledge to expert systems. They indicate that it is a mistake to believe that the techniques of knowledge acquisition are sufficiently defined to uncover the whole of an expert's knowledge. Polanyi notes that there is an "act of tacit inference" that differs sharply from conclusions based on induction or deduction. Tacit inference is integrative, bringing together ideas not in the style of inductive or deductive reasoning but in an intentional way that draws conclusions as a part of an expert's consciousness.

Polanyi addresses the problem of knowledge transfer from an epistemological perspective:

> ...even a writer like Kant, so powerfully bent on strictly determining the rules of pure reason, occasionally admitted that into all acts of judgement there enters, and must enter, a personal decision which cannot be accounted for by any rules. Kant says that no system of rules can prescribe the procedure by which the rules themselves are to be applied. There is an ultimate agency which, unfettered by any explicit rules, decides on the subsumption of a particular instance under any general rule or a general concept. And of this agency Kant says only that it 'is what constitutes our so called mother-wit'... he declares that this faculty, indispensable to the exercise of any judgement, is quite inscrutable ... the way our intelligence forms and applies the schema of a class to particulars 'is a skill so deeply hidden in the human soul that we shall hardly guess the secret trick that Nature here employs'...[70]

From this we see an inherent difficulty in knowledge acquisition and modeling. With the epistemological problems of machine intelligence in general and expert systems specifically, validity and reliability become key issues. It is our contention that a need exists for a strong theory of action, differing from the engineering approach, which takes precedence over a theory of knowledge and serves as a foundation for expert systems. Most contemporary philosophic systems have such a theory of action, commonly referred to as "praxis".[71] It is our contention that a distinction must be made between "praxis" and "poesis". "Praxis" is activity that has its goal within itself. Its aim is not the production of artifacts but rather the formation, internally, of particular activities in a certain way. "Poesis", or production, is activity whose aim is to bring something into existence distinct from the method of production itself.[72] The notion of praxis is an important concept for expert systems, for while the knowledge engineer is concerned with poesis, or the production of an algorithm, the expert is concerned with the internalized process.

Although there is an etymological similarity between praxis and practice, praxis is not a mundane bread-and-butter activity. Praxis includes the complex of issues relating to intentions, motives, purpose, reasons, and teleological explanations. Praxis in the context of expert systems is related to the essence of what decision making is; its ends are not necessarily those of production but rather are an attempt to weigh internally what one values — hence a connection with the notion of root metaphor.

AN EPISTEMOLOGICAL TYPOLOGY OF DECISION AID TECHNOLOGIES

We believe that the popularity of decision-aid technologies, coupled with the power they possess for problem solving in information-oriented systems, requires careful philosophic scrutiny. In addition, we feel that the use of metaphors can be easily incorporated into philosophic analysis as an aid to comprehension. Table 1 compares statistics, operations research, and expert systems and develops an epistemological typology to differentiate the attributes of these three decision technologies. Included in this typology are the activating root metaphors; the means and ends employed by a particular decision aid; the mode or manner used by the decision aid in its inquiry; its primary logic or inquiry; and finally, a metaphor that describes each of the decision aids in terms of syntactical mood, indicating its state of action.

The purpose of statistics is to develop, through comparison, a description of reality. Its manner or mode of inquiry is nominal, i.e. it attempts to name or categorize reality through a process of partitioning variance, developing taxonomies, and showing whether differences between or within subjects or groups are significant.

Table 1. An Epistemological Typology of Decision Support Systems Technologies

Decision Aid	Root Metaphor	Means	Ends	Mode of Enquiry	Synactical Mood	Logic
Statistics	mechanism	*a posteriori*	description	nominal	indicative	inductive
Operations research	formism	*a priori*	prescription	normative	subjunctive	deductive
Expert systems	contextualism	praxis	rescription	consultative	imperative	tacit

Statistics therefore speaks to us in a syntactical mood that resembles the indicative; it addresses what and how things are. Statistics as a decision aid has its foundations in the mechanistic root metaphor and can generally be assumed to be the principle tool of positivism. Statistics is an inferential method of inquiry: using inductive logic, it targets a series of experiential or *a posteriori* observations as the principle means to achieve its descriptive ends. This approach is most successful when we are concerned with describing the world as is, without judgement.

Operations research, working from an *a priori* means has as its ends prescription, that is, how can reality be improved. It is therefore normative and speaks from the subjunctive mood — "... if the world were as it ought to be it would look like this".[73] Techniques such as linear programming and decision analysis prescribe to the situation in an attempt to make improvements or to demonstrate the effects of alternative decisions. Operations research uses a predominantly deductive logic, such as probability theory, and has its foundations in idealism and the formistic root metaphor. Its greatest value is in providing the guidance for situations where human intuition is fallible.

An expert system achieves its ends through an interrelationship between the praxis of the expert and the poesis of the knowledge engineer. Its goal is to achieve rescription, that is, an act of reinterpretation (literally it attempts to rewrite an expert's decision process). Its mode of inquiry and syntactical mood address its intention to mimic a decision process through capturing how an expert makes a decision. An expert system is consultative; the interaction between the knowledge engineer and the expert represents a mode of inquiry that is dependent upon precise communication of knowledge and intention between them to establish the rules for a decision situation. Due to this mode of inquiry, an expert system "speaks" in the imperative mood.

Unlike the classical logic of statistics and operations research, which use induction and deduction as tools to achieve their ends, an expert system relies on a type of tacit inference. An expert system does employ inductive and deductive reasoning, but it also recognizes and attempts to replicate the prelogical, intentional, and intangible inferences of a decision maker (analogous to our previous example of the tacit knowledge needed to ride a bicycle). It serves as a reflective force between the knower and the known, and because of this intuitive dimension, it exists prior to logical influence. The principal philosophical foundation of expert systems is the contextualistic root metaphor; it is inherently pragmatic and open-ended. Expert systems have been found to be most useful in situations where an individual has extensive experience with a decision situation and can express this knowledge in a logical way.

A DIAGNOSTIC PROCEDURE FOR DECISION SUPPORT SYSTEMS

We believe that a diagnostic procedure for decision support systems should be developed based on the typology presented in Table 1. Although it is difficult to define an individual decision maker's root metaphor, the means, ends, syntactical mood, mode of inquiry, and logical base of a decision aid are more easily recognizable. Figure 2 outlines a possible diagnostic procedure to follow employing the typology in Table 1.

The decision making process starts from the perception of a problem; if a problem is not perceived, there is no need for a decision to be made. Once the problem has been recognized, the next step is to formulate the problem statement and definition. This constitutes the "meta-decision problem", i.e. the problem of a problem. The problem statement and problem definition are the most crucial aspects of the decision-making process. This stage in the decision process defines the constraints and boundaries of a problem and gives the first indications of the alternatives available.

To assess whether the problem has been misconceived, an adequate definition must include the problem structure. In this situation, we suggest that the decision maker employs the concepts of Table 1 and asks the following questions.

(1)   In what mood of syntax is the problem stated? Is there an implicit "ought to be" that will indicate the subjunctive, or does the problem seem to be in the indicative or imperative?

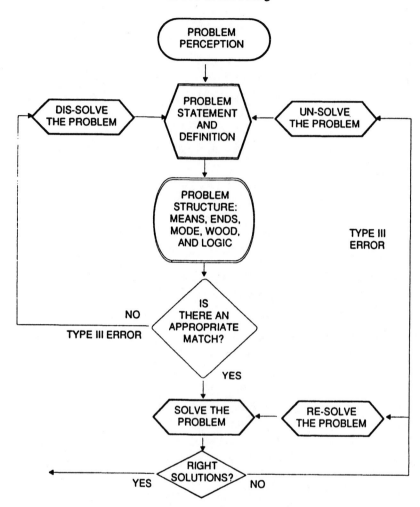

Fig. 2. Error diagnostic procedure. (Adapted from Dunn.[18])

(2)   What means will be employed to measure the problem? Will it be theoretical, i.e. will it be based on some *a priori* view of the world (operations research)? Will data be collected and analyzed from an *a posteriori* stance (statistics)? Or will it be an attempt to mimic some internal processing of a decision maker (expert system)?

(3)   What is the logical flow of the discovery process? General to specific? Specific to general? Or is there an unascertainable tacit dimension to the situations logic?

(4)   Is the mode of inquiry nominal, normative, or consultative?

(5)   Finally what are the ends of the decision problem? Are they descriptive, prescriptive, or rescriptive?

These steps result in a determination of the "correctness" of the problem definition. If we cannot match the decision aid technology to the typology in Table 1, we may have encountered a Type III error. If so, the problem should be "dis-solved",[74] and the problem-structuring process should be done again.

After the problem has been solved, the question becomes, "Do we have the right solution?" If the answer is "no", a distinction must be made as to whether a Type I, II, or III error has occurred. If the decision maker perceives it to be a Type I or II error, the problem can be "re-solved". If the problem solution is incorrect and the decision maker has ascertained that no Type I or II error has been committed, then a Type III error should be suspected and the problem "un-solved".[75] Another iteration through the problem definition stage should then be performed. We recognize that not every decision situation will be congruent with the typology. However, we do submit that forcing the decision maker to assess any lack of congruence will help reduce Type III errors.

CONCLUSION

It is our conviction that the proliferation of decision-support software and its user-friendly environment, while generally beneficial, may result in Type III errors in the hands of naive users. In addition, we agree with Raiffa[76] that subtle errors can be caused by poor communication between decision makers and analysts. Due to the contingent and variable nature of decision situations, it is imperative that each situation be considered unique. An indispensable need exists to test the congruence between (1) the philosophical foundation of a decision aid, (2) the root metaphor of the person using the decision aid, and (3) the decision situation to see whether a particular decision aid is appropriate.

The typology and diagnostic procedures we propose can be invaluable aids in testing that congruence. It is our hope that this chapter stimulates inquiry into the philosophical underpinnings of management sciences and encourages proper use of the exciting and productive decision aids available to all of us.

NOTES AND REFERENCES

1. Mulvey, J.M. and Blount-White, S.E., "Computers in the Government: Modeling and Policy Design", Dep. Civil Engineering, Princeton University, Princeton, NJ, 1986.

2. Churchman, C.W., *Prediction and Optimal Decision: Philosophical Issues of a Science of Values.* Englewood Cliffs, NJ: Prentice-Hall, 1961, and *The Design of Inquiring Systems.* New York: Basic Books, 1971.

3. See Dennett, D.C., *Brainstorms: Philosophical Essays on Mind and Psychology.* Cambridge, MA: MIT Press, 1987; H. Putnam, "The Mental Life of Some Machines", in *Intentionality, Minds and Perception*, H. Castaneda, ed., Detroit, MI: Wayne State University Press, 1967; and L. Winner, "Complexity and the Limits of Human Understanding", in *Organized Social Complexity*, T. R. LaPorte, ed. Princeton, NJ: University Press, 1975.

4. See McCarthy, J., "Epistemological Problems of Artificial Intelligence", in *Proceedings 5th International Joint Conference Artificial Intelligence.* Cambridge, MA: Massachusetts Institute of Technology, 1977, and J. McCarthy and P. J. Hayes, "Some Philosophic Problems from the Standpoint of Artificial Intelligence", in *Machine Intell*, B. Meltzer and D. Michie, eds. Edinburgh, UK: Edinburgh University Press, 1969.

5. Ginzberg, M.J. and Stohr, E.A., "Decision Support Systems: Issues and Perspectives", in *Decision Support Systems*, J. Ginzberg, W. Reitman, and E. A. Stohr, eds. Amsterdam, The Netherlands: North Holland, 8, 1982.

6. Bodily, S.E., "Spreadsheet Modeling as a Stepping Stone", *Interfaces*, **16**, 34–52, 1986.

7. Savage, L.J., *The Foundations of Statistics.* New York: Wiley, 1954.

8. Churchman, C.W., *op. cit.*

9. See Alter, S., *Decision Support Systems: Current Practice and Continuing Challenge.* Reading, MA: Addison-Wesley, 1980; R. W. Blanning, "The functions of a decision support system", *Information Management*, **2**, 1979, 87–93; M. W. Hurley and W. A. Wallace, "Expert Systems as Decision Aids for Public Managers: An Assessment of the Technology and Prototyping as a Design Strategy", *Public Administration Rev.*, **46**, 1986, 563–571; J. D. C. Little, "Models and Managers: The Concept of a Decision Calculus", *Management Science*, **16**, 1970, 5466–5485; R. H. Sprague, "A Framework for the Development of Decision Support Systems", *MIS Quarterly*, **4**, 1980, 1–26; D. Young, *Brief Usability Survey of Operations Research Applications Software for Decision Support Systems.* Atlanta, GA: Georgia Institute of Technology, 1978; and W. Zachary, "A Cognitively Based Functional Taxonomy of Decision Support Techniques", *Human-Computer Interaction*, **2**, 1986, 25–63.

10. Winner, L., *op. cit.*, 65.

11. Sprague, R.H., *op. cit.*

12. Raiffa, H., *Decision Analyses: Introductory Lectures on Choices under Uncertainty.* Reading, MA: Addison-Wesley, 1970.

13. *Ibid.*, 265.
14. Davis, G.B. and Olsen, M.H., *Management Information Systems: Conceptual Foundations, Structure and Development.* New York: McGraw-Hill, 1985.
15. Mitroff, I.I. and Turoff, M., "On Measuring the Conceptual Errors in Large Scale Social Experiments: The Future as Decision", *Technol. Forecasting and Social Change*, **6**, 1974, 389–402.
16. Simon, H.A., "Some Strategic Considerations in the Construction of Social Science Models", in *Mathematical Thinking in the Social Sciences*, P. Lazarsfeld, ed. Glencoe, IL: Free Press, 1954.
17. Morris, W.T., *Management Science in Action.* Homewood, IL: Richard D. Irwin, 1963, and "On the Art of Modeling", *Management Science*, **13**, 707–716, 1967.
18. Mulvey, J.M., "Strategies in Modeling: A Personnel Scheduling Example", *Interfaces*, **9**, 66–76, 1979 and "Computer Modeling for State Use: Implications for Professional Responsibility", *IEEE Technol. Society Magazine*, **2**, 3–8, 1983.
19. Morris, W.T., *Management for Action: Psycho-Technical Decision Making.* Reston, VA: Reston, 1972.
20. Luijpen, W.A., *Existential Phenomenology.* Pittsburgh, PA: Duquesne University Press, 1960.
21. Descartes, R., *Meditations on First Philosophy.* Indianapolis, IN: Hackett, 1979.
22. Dancy, J., *An Introduction to Contemporary Epistemology.* New York: Basil Blackwell, 1985, and W. J. Luijpen, *op. cit.*
23. Dancy, J., *op. cit.*
24. Helmer, O. and Rescher, N., "On the Epistemology of the Inexact Sciences", *Management Science*, **6**, 25–52, 1959.
25. Churchman, C.W., *op. cit.*
26. Mason, R.O. and Mitroff, I.I., "A Program for Research on Management Information Systems", *Management Science*, **19**, 475–487, 1973.
27. Mitroff, I.I. and Sagasti, F., "Epistemology as General Systems Theory: An Approach to the Design of Complex Decision-making Experiments", *Phil. Social Sci.*, **3**, 117–134, 1973.
28. Mitroff, I.I., Barabba, V.P. and Kilmann, R.H., "The Application of Behavioral and Philosophic Technologies to Strategic Planning: A Case Study of a Large Federal Agency", *Management Science*, **24**, 44–58, 1977.
29. Helmer O. and Rescher, No., *op. cit.*
30. Von Bertalanffy, L., *General System Theory: Foundations, Development, Applications.* New York: Braziller, 48, 1968.
31. Husserl, E., *The Phenomenology of Internal Time Consciousness.* Bloomington, IN: Indiana University Press, 1964.
32. Brown, R.H., *A Poetic for Sociology.* Cambridge, UK: University Press, 1977.
33. Pepper, S.C., *World Hypotheses.* Los Angeles, CA: University of California Press, 1942.
34. Churchman, C.W., *The Design of Inquiring Systems, op. cit.*
35. *Ibid.*

36. Mitroff, I.I. and Kilmann, R.H., "On Evaluating Scientific Research: The Contribution of the Psychology of Science", *Technol. Forecasting and Social Change*, **6**, 389–402, 1982.

37. Dunn, W.N., "Methods of the Second Type: Coping with the Wilderness of Conventional Policy Analysis", *Policy Studies Rev.*, in press.

38. Giere, R.N., *Understanding Scientific Reasoning*. New York: Holt, Rinehart and Winston, 1984.

39. Box, G.E.P., "An Apology for Ecumenism in Statistics", in *Scientific Inference, Data Analysis, and Robustness*, G.E.P. Box, T. Leonard, and C. Wu, eds. New York: Academic, 51–84, 1983.

40. Leonard, T., "Some Philosophies of Inference and Modeling", in *Scientific Inference, Data Analysis, and Robustness*, G. E. P. Box, T. Leonard, and C. Wu, eds. New York: Academic, 9–24, 1983.

41. Savage, L.J., *op. cit.*

42. Barnett, V., *Comparative Statistical Inference*. New York: Wiley, 1982.

43. *Ibid.*, 6.

44. Giere, R.N., *op. cit.*

45. Churchman, C.W., *The Design of Inquiring Systems, op. cit.*

46. *Ibid.*

47. Brown, R.H., *op. cit.*

48. Kuhn, T., *The Structure of Scientific Revolutions*, 2nd edn. Chicago, IL: University of Chicago Press, 1970.

49. Hofstadter, D.R., *Godel, Escher, Bach: An Eternal Golden Braid*, New York: Basic Books, 1979.

50. Churchman, C.W., *The Design of Inquiring Systems, op. cit.*

51. Rothenberg, H., *The Emerging Goddess: The Creative Process in Art, Science, and Other Fields*. Chicago, IL: University of Chicago Press, 1979.

52. Von Bertalanffy, L., *op. cit.*

53. Turing, A.M., "Computing Machinery and Intelligence", *Mind*, **59**, 433—460, 1950.

54. Sutherland, J.W., "Assessing the Artificial Intelligence Contribution to Decision Technology", *IEEE Trans. Syst. Man Cybern.*, **SMC-16**, 3–20, 1986.

55. *Ibid.*

56. Munevar, G., *Radical Knowledge: A Philosophical Inquiry Into the Nature and Limits of Science*. Indianapolis, IN: Hackett, 1981.

57. See Lucas, J., "Minds, Machines and Godel", in *The Modeling of Mind*, (K. Sayre and F. Crosson eds.) Notre Dame, IN: University of Notre Dame Press, 1963, and L. Winner, *op. cit.*

58. Waterman, D.A., *A Guide to Expert Systems*. Reading, MA: Addison-Wesley, 1986.

59. *Ibid.*

60. Johnson, P.E., "The Expert Mind: A New Challenge for the Information Scientist", in *Beyond Productivity*, T. Bemelmans, ed. Amsterdam, The

Netherlands: North Holland, 1983.

61. Helmer O. and Rescher, N., *op. cit.*
62. Mitroff, I.I. and Kilmann, R.H., *op. cit.*
63. Janis, I.L. and Mann, L., *Decision Making*. New York: Macmillan, 1977.
64. Morris, W.T., *op. cit.*
65. Festinger, L., *A Theory of Cognitive Dissonance*. Evanston, IL: Row Peterson, 1957.
66. Polanyi, M., *The Tacit Dimension*. Garden City, NY: Doubleday, 1966.
67. *Ibid.*
68. *Ibid.*, 7.
69. See Collins, H.M., "The TEA Set: Tacit Knowledge and Scientific Networks", *Sci. Studies*, **4**, 165-186, 1974 and H. M. Collins, R. H. Green, and R. C. Draper, *Changing Order: Replication and Induction in Scientific Practice*. Beverly Hills, CA: Sage, 1985.
70. Polanyi, M., *Personal Knowledge*. Chicago, IL: University of Chicago Press, 207, 1958.
71. Bernstein, R.J., *Praxis and Action*. Philadelphia, PA: University of Pennsylvania, 1971.
72. Runes, D.D., *Dictionary of Philosophy*. Totowa, NJ: Littlefield, Adams, 1962.
73. Churchman, C.W., *Prediction and Optimal Decision, op. cit.*
74. Dunn, W.N., *op. cit.*
75. *Ibid.*
76. Raiffa, H., *op. cit.*
77. Mitroff, I.I. and Betz, F., "Dialectical Decision Theory: A Meta-theory of Decision-making", *Management Science*, **19**, 11–24, 1972.
78. Yazdani, M. and Narayanan, A., *Artificial Intelligence: Human Effects*. New York: Wiley, 1984.

Chapter 2

Part 3. Models in the Public Sector: Success, Failure and Ethical Behavior

John M. Mulvey

INTRODUCTION

Computers increasingly affect our everyday life in unexpected ways. Not only do computers track our credit card balances and mortgage payments, they also make decisions by means of complex mathematical models. We call this technology "computerized decision procedures", or CDP's for short.

The United States Federal Government is a prime source for the development of computer planning models. Almost every large federal agency has an equally large series of CDP's on which to base its policy recommendations. Several typical examples are discussed in this chapter.

Many of these techniques are poorly understood by the general public and even by the users who interact with the CDP's. The lack of understanding is not unexpected, given the complexities of many CDP's. Also, most high-level politicians in the United States are not trained in quantitative reasoning and mathematics. Yet the same individuals must deal with CDP's on a recurring basis. Congressional support staff and organizations such as the Office of Technology Assessment are employed in the task of translating the technical issues into "plain English" for the decision makers.

Many of these studies are done under extreme time pressure and with limited resources. The "hot" issues often prompt Congress to act before there is time to prepare in-depth analyses. Besides, members of Congress are hesitant to accept complex answers for our societal problems. It is much easier for them to sell simple solutions to their constituents.

Combining all of these elements gives rise to the potential for misuse of the CDP technology. (We are all too aware that high technology presents a double-edged sword.) Generally, we believe that the abuses are done innocently enough, but there is little to stop a determined analyst from "adjusting" a CDP in order to generate desired conclusions and from getting away with the subterfuge. The professional societies have paid almost no attention to possible unethical conduct, and they have virtually no power to prevent an analyst from practicing his trade no matter how dishonest he might have been.

The topic of ethics in modeling must, by necessity, deal with the notion of harm. The actions of one group must somehow negatively affect the well being of others. Codes of ethical conduct represent attempts to reduce the harm that might occur when professionals carry out their business. However, to date, public acceptance of the notion that a computer program can do actual physical harm without human intervention is unlikely to take place for a decade or two. (The increasing use of computers to fly aircraft may accelerate this projection, however.)

As far as CDP's are concerned, the definition of harm should include indirect effects. For example, if you were not promoted to a new job or were not assigned to a hospital for your medical residency because a CDP did not feel that you were qualified enough, you might make the case that you have been harmed. However, if a CDP was part of a debate on tax policy and lawmakers decided that capital gains taxes should be decreased in order to reduce the national debt, the argument of harm makes a less convincing case. After all, our representatives take ultimate responsibility for decisions, not the computers.

While of course this position is true, the use of a CDP can have an enormous impact on the deliberations of a policy debate. The quality of the decisions will depend upon the information provided, and CDP's are transmitters (and transformers) of information. The politicians' advisers and consultants must share some of the responsibility for the politicians' decisions. The responsibility is greatest when the topic involves highly technical issues in which the politicians are poorly trained.

It is in this spirit that the topic of ethics is discussed. There is a grave potential for harm to the public or to certain groups in the public, but the harm can be indirect and possibly diffuse — affecting many people, perhaps in the future. However, the issues raised will become increasingly important in the decades ahead as the world becomes more technologically complex and interrelated.

FOUR HIGHLY SUCCESSFUL CDP'S

Each of the four planning systems discussed below has been used successfully for a number of years. These CDP's come from the public sector and are part of normal operations. It is unlikely, therefore, that any of the systems could be replaced by a human, since the various groups have developed a strong dependence on the computer. Most people associated with these systems seem to feel that their system's recommendations are generally acceptable. Altogether, little enthusiasm exists for any formal evaluation of the current procedures. As far as I have been able to tell, the original designers of each system are no longer part of the process.

Despite the obvious interest in maintaining an operational CDP, I believe that periodic reviews of the design and the implementation of a CDP should be made for the purpose of determining if the system is still the most appropriate for servicing its current tasks. In other words, the CDP may have outlived its usefulness — given changes in hardware and software, the availability of new databases, modifications in the objectives of the policy makers, and so forth. These reviews should determine if the current CDP is the most appropriate.

At the same time, since these systems are part of operations, there is a strong reluctance to allow for their formal evaluation. Many people disapprove of the idea of evaluating their "personal" computer programs. And what should be done with the current system if the evaluators decide that the current system needs replacing? Should any CDP's be condemned?

And if so, what should be done with the decision-making process while a new system is being built? I'm afraid that the answers to these important questions have not been found.

Perhaps CDP's have a life of their own. Partially because of the issues raised above, the replacement of large CDP's rarely takes place. The costs of major redesign in terms of actual cash outflows plus the opportunity costs can be too high to undertake. Of course, these systems are modified in an incremental fashion over time. The temptation to tamper with the CDP is large. Remember that the amount of work to change a CDP can be very small indeed. It might be simply the addition of a new function or feature, such as changing the objective function to handle piecewise linear costs. Or it might entail a change in the planning horizon, for example, using a tactical planning model for addressing a strategic issue. Or the CDP might be set up to solve a much different problem. The variations are endless.

As you read the review of the four successful CDP's, give some thought to the problem of measuring the ultimate success of a computerized decision procedure. What defines success in these instances?

Office of Tax Analysis

CDP's have become an indispensable part of many political debates. During debate on the 1986 Tax Reform Act, for example, the politicians decided that any changes in the tax code would be "revenue neutral", meaning that the overall net impact on tax revenues would be essentially zero. The gains and losses to various constituents would balance out. Instead of arguing about unresolvable issues, such as regarding the long-term impacts of tax changes on behavior and therefore revenues (remember that the tax debates in the early 1980s turned on whether or not total revenues would increase or decrease if the tax rates were drastically lowered), the politicians decided to use the Office of Tax Analysis (OTA) and its various computerized models. In effect, the debate was shaped by the Office of Tax Analysis' CDP's. Revenue neutrality began to mean "as predicted by the models", rather than actual revenue neutrality. Of course, it would have been impossible to measure the actual revenue impacts at the time of the debate.

The use of CDP's to generate calculations and predictions for use in political debates will undoubtedly increase. Our complex society demands that the important issues be considered when policy decisions are made. As mentioned, these decisions are often rendered under extreme time pressure, due to the nature of Washington politics. For instance, the return of the tax debate every couple of years is as certain as the return of the new moon each month. Yet the serious political haggling does not get started until several months (or even weeks) before the bill emerges from the Joint Committee on Taxation. The only effective way to accommodate this process is to prepare CDP's for estimating the distributional impacts of the deliberations. The OTA staff has seen many cycles and has done a first-rate job in preparing the tax calculators (as they are called).

A final point on the OTA example is in order. These models are extremely large in terms of the numbers of variables. After all, estimating distributional impacts is a complex task. Who knows what Congress will be interested in? They might wish to measure the impact on minority women in New Jersey who have an adjusted gross income over US $50,000. Or they might want to focus on the elderly poor in Colorado. To accommodate these requests requires a large and thoroughly developed database. The OTA spends several years getting this data together — again using a number of CDP's to clean up and enhance the information.[1] The result is called a microdatabase.

FAA Traffic Control

Large-scale CDP's affect our everyday life. As an example, the Federal

Aeronautics Administration (FAA) employs a real-time computer system for predicting traffic congestion at the busiest United States airports.[2] Once a potential bottleneck is discovered, a CDP is called in to delay flights headed for the congestion. Every effort is made to delay flights on the ground at the gate rather than requiring airplanes to go into holding patterns. Readers who are old enough may remember circling endlessly over Lake Michigan while waiting for an open landing slot at the perennially busy O'Hare field near Chicago, Illinois. The FAA flow-control CDP has largely eliminated these expensive delays. Of course, the delays now occur at the gate, due to the explosive growth in air travel over the past decade. The FAA model "decides" which planes will be delayed and by how much, based on its forecasts of future traffic at the busiest airports.

Forest Service 150-Year Planning

Since the introduction of cost/benefit analysis in the 1930s, the federal government has depended upon mathematical formulas for assisting in the task of allocating scarce resources. These same procedures are generally required when large public works are decided upon; the goal is to undertake projects that create greater benefits than losses. Naturally, it is a matter of some importance who decides what the benefits and costs are and how they are measured. Also the long time spans may result in intergenerational conflicts.

More recently, CDP's have assumed the task of calculating costs and benefits for proposed policy decisions. The United States Forest Service has led the way. Its forest planning system (FORPLAN) — set up in response to the National Forest Management Act of 1976 — requires tens of thousands of decision variables. Massive amounts of computer resources are needed to solve the resulting large-scale linear program. The aim is to maximize the "public welfare" over the next 100–150 years. Constraints are imposed on the system by means of algebraic equations. For instance, the model ensures that timber harvests will increase over time without depleting the forest base. Other goals are considered, such as setting aside land for grazing animals. Again, there is considerable controversy regarding methods for measuring the large number of factors along with the process for combining the various attributes into a single linear objective — a public welfare function. In addition, the complexity of the societal issues (tradeoffs between the environment and economic good, for example) causes concern among some. However, these CDP's are in active use.

Personnel Planning (Army and Medical Matching)

The service economy is not immune to the growth of CDP's. A noteworthy example involves the assignment of people to jobs in large organizations. For many years, the military has routinely employed planning systems for predicting the imbalances in supply and demand of its personnel.[3] These systems assist in the task of hiring and promotions, among others.

A CDP with a similar purpose is the National Residency and Internship Matching Program (NRIMP), or the Medical Match, for short. This system begins with preferences given by fourth-year medical students for hospital assignments and preferences given by hospitals for the interviewed students. Over 14,000 students participate annually. Anyone joining the matching CDP must abide by the results. It is a compulsory system for the hospital and students.

The input data for the Match CDP is easy to describe: (1) a ranked list of acceptable hospitals from each student, and (2) a ranked list of acceptable students from each hospital. These ranks are based on ordinal preferences, i.e. (1, 2, 3, ...), without ties. It is much more difficult to discover the overall objectives of the Match CDP. The computer performs a series of steps in which students are assigned to hospitals. These rules are embedded in an algorithm known as the Student-first procedure. Generally, the process is to first "match" the top-rank students to the top-rank hospitals. Once these assignments are made, the next ranks move up and the process repeats itself. Occasionally, students are reassigned to higher ranking hospitals. One might envision this process as a series of queues in which students stand in line in front of their chosen hospitals, waiting to be chosen. The CDP makes the choices according to the programmed rules.

EVALUATING MODELING SUCCESS OR FAILURE

The mark of a successful modeling project can be measured in several ways. A happy client who is willing to continue to use the CDP and fund future modeling projects is an obvious requirement. The four systems discussed in the previous section certainly fulfill this criterion. For a permanent impact, however, there should also be a relatively objective evaluation of the advantages of the CDP. For instance, the businessman will want to see how much the system is saving in terms of greater efficiency, perhaps, or in generating new revenues. The savings should substantially outweigh the costs of the CDP, in terms of installation and maintenance costs. A periodic appraisal is essential in order to determine if the system is performing properly.

The evaluation issues become more complex when the CDP involves the public sector. Often, mixed motives underlie a CDP's implementation. Take the case of the Forest Service Planning System. The forest planners want to design a procedure that is somehow "fair" to the various constituents: The timber industry would like to increase its production of felled trees so that profits will increase. The local workforce needs the jobs. The environmental groups want to see clear cutting stopped. Congress would like to see exports increased. And so on.

Each of these sides will often measure the success of FORPLAN in terms of how well it achieves its objectives. Did the system eliminate clear cutting? How much timber production is allowable? What will happen to the economic conditions of the Pacific Northwest? How much recreation is possible? FORPLAN's recommendations may be judged in the light of its recommendations.

Furthermore, many CDP's require forecasting subsystems that reach many decades into the future. The large uncertainties present are generally handled through scenario (or sensitivity) analysis; a consideration of the full complexity of the uncertainties within the context of a stochastic program or similar approach is prohibitively expensive. Unfortunately, the forecasts are rarely calibrated due to the long planning periods. In effect, one would have to wait 50 to 150 years before "knowing" if the forecast is accurate. The use of probability-based forecasts require even more time for their calibration. Thus, a purely scientific approach to measuring the success of a CDP is unlikely to be practical for some time.

We are thereby left with the indirect benefits of CDP's as indicators of modeling success. These include: (1) the judgments of the parties about the system's usefulness; (2) the inevitable improvements in data that result when a computer model is introduced; (3) the employment of Operations Researchers; (4) the hope that the political issues will be clarified by a computerized model; (5) the ease of using the resulting CDP so that just about anyone can gain an appreciation of capabilities.

Unfortunately, the most thorny societal problems are seldom amenable to a single systematic approach. Diverging groups will often develop CDP's that have much different structures and that generate much different recommendations. It is as though the models reflect the souls but not the bodies of the developers.

The professional societies seemingly have not recognized the usefulness of alternative models and systems, since few ideas exist about how competing CDP's should be compared. If pressed, most OR professionals fall back on the scientific method as the key to picking "the correct" model. However, the ultimate success of a public-sector CDP hinges on the aforementioned indirect benefits, and these are much more difficult to ascertain by means of quantitative processes.

The model selection process has an important ethical component. Choices must be made regarding the elements to be included in the model. Typically, the availability of data and the project budget dictates many of the choices. But the elements that have hard, measurable features — such as the amount of lumber in board feet — can drive out the less quantitative aspects, such as aesthetics. Of course, attempts can be made to measure all of the concerns. FORPLAN takes into account animal-grazing days as part of its objectives. It is much more difficult to compute the worth of an animal-grazing day. And what about the number of butterflies? Do we want to count the annual number of contacts between butterflies and humans in each of the National Forests? Admittedly, many aspects of the societal problem are hard to incorporate in a CDP. Sound judgment on the part of the users must be relied upon to overcome these inherent limitations of CDP's.

The success of a CDP must take into account the political process along with the objective determinants, such as cost savings. Remember that CDP's have the potential to be used as political weapons, rather than as scientific, objective tools for inquiry. Ethical dilemmas must also be studied. The elements of a successful CDP must deal with these universal human issues and cannot be left solely with the technicians — however well trained they are.

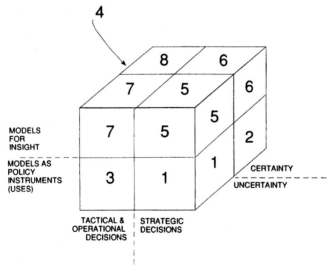

Fig. 1. The modeling cube.

As a final point, we note that CDP's have a variety of purposes. Figure 1 categorizes these into eight broad classes, along three dimensions. The dimensions are as follows: (A) Models for Insight, or Models as Policy Instruments; (B) Tactical and Operational, or Strategic; and (C) Certainty, or Uncertainties.

The definition of a CDP's success will depend upon which category it fulfills at the time of the evaluation. For instance, the success of a system in class #8 (Models for Insight, Tactical Decisions, and Certainty) can be judged by asking the user how much insight he or she got. The CDP's in the lower numbered classes will require considerably more work in order to tell if they are truly a success. All CDP's should not be lumped into a single set of rules for determining modeling success.

RECOMMENDATIONS

There is a tendency to develop either too much or too little enthusiasm when it comes to the topic of computers. This inclination is particularly evident for computer software tools. Take the loyalties analysts possess for particular operating systems. One has only to ask a computer expert to trade his Apple-Macintosh for a Unix-workstation in order to see the strength of this brand loyalty.

A similar situation occurs when it comes to operations researchers and their favorite models. One analyst would never use Monte-Carlo simulation over mathematical programming. Another analyst would always make the reverse judgment. Modelers seem to bring their own perspective to the modeling process.

There is nothing inherently wrong with these biases, of course, so long as there is a valid attempt to compare the various alternative model representations. The pros and cons of the competing models ought to be evaluated systematically. Unfortunately, the comparative analysis is rarely performed!

To overcome these biases, my first recommendation is to improve the modeling process by evaluating the major competing approaches, especially for public-sector applications. Much research is needed in this regard. The OR/MS profession has paid almost no attention to this issue. Models are generally presented in journals and at professional meetings as if there were only a single correct representation: the one presented. There is virtually no discussion about the choices that were made by the modelers or the tradeoffs encountered. Surely the model is a compromise between the competing objectives of model realism, computational and informational costs, and user friendliness.[6] These choices affect the model's final design and therefore should be made available to the various parties involved. The evaluation should be conducted in an open fashion.

Preventing this open comparative analysis is the desire by many analysts to employ their CDP as a form of persuasion, rather than as an objective tool of scientific inquiry. Tightly controlled information provides strength in a political setting and ultimately job security for the model developer. Also, a complex model is much harder to duplicate by competitors and thus provides a natural barrier to entry by competitors. The Defense Department is notorious in its use of massive CDP's and the classified data needed to run these behemoths. Only the big defense contractors are able to mount a serious challenge when requests-for-proposals (RFP's) are sent out. Thus, many defense-related CDP's have been run and rerun for literally decades without a major redesign. Meanwhile, the systems engineering group is able to generate millions of dollars in consulting fees. These CDP's rarely see the light of day, for obvious reasons.

The remaining recommendations are presented from two vantage points — a pessimistic and an optimistic view. Given that my livelihood depends upon the use of large CDP's, I am anxious to see this technology grow and prosper. At the same time, I would especially like to see this technology used properly for improving the public good. Our society's future well being may depend on it.

A Pessimistic View

The main difficulty with the notion of establishing a code of ethical conduct for professionals involves the implementation of any recommendations. What mechanisms are available for enforcing a breach of ethics? Is there a disciplinary group that can first render judgments and the presence of unethical behavior and then decide upon a just penalty?

Other professions have a tighter definition of professional success (or more importantly failure) than does operations research. For example in medicine, was harm done to a patient? Without this shared value, it is difficult to make a tough decision about professional misconduct. Perhaps this explains why no Operations Researcher has ever lost his or her membership in the OR Society. There are simply too many motives for carrying out a modeling project for there to be a single overriding value.

Another factor on the negative side of the ledger is the strong financial incentive to carry out a "closed evaluation". The open-door policy can be thought of as a way to give the other side ammunition so that they can refute your analysis. Who would be stupid enough to work for the automobile companies and pass out the raw data for their safety studies? Regardless of a study's outcome, there would be a tradeoff between cost and safety — as measured in stark numbers. A prosecuting attorney for someone who was killed in a crash involving your car would have a field day with this analysis. The attorney would simply show that the cold, cruel auto manufacturer was gaining a bit of extra profit at the price of someone's life. We all know that human life cannot be priced.

This same issue occurs in the public sector. The FAA will never measure the safety of individual elements in the airline travel system and compare it against cost efficiency in a formal way. Instead, the FAA repeatedly states that the system is deemed safe. Clearly, the FAA is saying that the system is safe enough — as measured by the FAA. Any models that take up the risk/cost tradeoff are deemed unacceptable. In this environment, the use of CDP's for analyzing risks and costs must be done behind closed doors. Political expediency dictates.

An Optimistic View

Given the realities of United States political decision making, one might argue that no model is better than a complex CDP. However, at least in my opinion, there is a strong case to be made for using a CDP for assisting policy makers provided that certain safeguards are employed. This section looks at the problem from the positive perspective.

The case for a CDP hinges on the availability of sound and relevant data. A model is not better than the data on which it is based. In fact, a CDP can be thought of as a transformer of data. We have all felt the impact of the old expression "garbage in, garbage out". Therefore, when the modeling process requires that good data be generated, a positive benefit is encountered. Many other groups can use the data as it is collected. The Office of Tax Analysis model provides an ideal example of this secondary benefit.

Second, the modeling process educates the developers in a unique way. When each aspect of a complex problem is dealt with in a careful manner, there is much to be gained in terms of better understanding of the decision problem. For this reason, I believe that the policy makers must be considered an essential part of the model development phase of implementation. Policy makers who are brought in at the later stages will not be familiar with the design tradeoffs and the rationale for choosing the end design. The technical people must make the tough decisions in this environment. Policy makers thereby do not comprehend the limitations of the CDP.

The next advantage of a CDP involves the consistency that flows from using a repeatable process. Note that we are focusing primarily on operational CDP's, ones that are running hundreds or thousands of times per year. The mathematical nature of a CDP forces the process to recommend the identical output when fed with identical inputs. Humans are not as good at attaining this level of consistency; their tastes and habits change, for instance. Consistency over time ensures that equity occurs between the generations. Also, it is quite desirable to achieve consistency in certain fields, such as portfolio investment modeling, in which the goal is to handle risk/reward tradeoffs over time. Inconsistency in this field leads to erratic (and generally suboptimal) results. This accounts for the popularity of the computerized-program trading strategies on Wall Street. There can be much money made by using a CDP for financial planning.

Next, when a CDP is developed in an open fashion — not bound by the need to generate *the correct recommendation* — there is much to be gained in terms of new insights. Unexpected results are the hallmark of a successful CDP. One would not need a fancy computer system if all it did was confirm one's educated hunches. Why bother to spend so much money on an immobile parrot?

The recommendations of a CDP should be treated as the advice of an expert. It should be scrutinized from several standpoints. The reputation of the model makers must be evaluated, for example, to decide if they have the proper level of training in order to complete the assigned task. The developers' previous projects should be studied to see if they have been able to build workable CDP's. Are their models easy to understand and operate? What about the user's manuals and other forms of documentation? Is the system able to handle sensitivity analyses so that the policy analysts can gain an understanding of the relationship between the input parameters and the output? Will the modeler's organization stand behind the product for several years? What about porting to other computers as new machines become cheaper and more powerful? These and many other issues should be considered.

Since many CDP's are designed as planning tools, they must by necessity deal with the dimensions of time and uncertainty. The latter issue gives rise to many diverse approaches. Some CDP's simply replace the random variables with point forecasts (e.g. chance-constrained LP's). The forest planning model is an example. Other CDP's attempt to model uncertainty directly by including penalties for overage or underage or by designing a second stage in which recourse decisions are included. These models go by the name Stochastic Programming or Monte Carlo Simulation.

Another strategy is to evaluate risk as the variance in outcomes that are possible.[7] Regardless of the approach taken to uncertainty and time, the construction of a CDP forces the policy maker to articulate his or her "policy". He or she must be able to describe the objectives that are sought and the approaches for meeting these objectives. Thus, a side benefit of a CDP can be a system for evaluating if the policy is being achieved. President George Bush may speak eloquently and long about his desire to set up a Nationwide Drug Plan, but unless the goals are clearly stated (and in measurable terms), there will be much confusion about whether or not the country is better off over time. Of course, precision of this sort seems to be outside the domain of United States politicians. Ask yourself the question about the Reagan era: Are we better off now than in 1981? There are many dimensions to issues of this type; thus it is unlikely that a CDP will be able to provide a single answer to this question. However, a CDP might shed some light on the distributional issues, such as measuring the gap between the top decile and the bottom decile income. A wider gap means that income equity has decreased. A narrower gap means that equity has improved. However, even this issue has political overtones. Some might argue that a wider gap must occur when the total level of income rises — the trickle-down theory. In any event, a simple CDP would shed some light on these issues. But a complex CDP would be unable to get very far, since the modeler would need to develop a set of cause-and-effect relationships, and these would quickly be subject to much debate. Hence the simplicity of the Office of Tax Analysis models contributes to its success — everyone is able to understand the process.

Used-Car Salespeople and the Conduct of Operations Research

Imagine that you are faced with the task of addressing a convention of used car salespeople about the topic of ethical conduct. They are attempting to improve their public image. What might you tell the salespeople?

The foremost advice would be for the salespeople to "tell the truth" about the state of their vehicles. An objective and believable survey of a car's condition would do wonders for the reputation of the used-car sales profession.

This advice about truth telling is particularly important when it comes to areas that are not obvious to the consumer. For example, the condition of the tires is almost immediately evident to someone who bothers to look. A leaky transmission that has been temporarily plugged, however, is another matter. It will often take a trained mechanic to discover all of the surprises that might befall a typical used-car buyer. Caveat emptor reigns supreme.

A second adage for the used-car sales profession might be: don't cause harm by selling an unsafe vehicle to an unwary buyer. The difference between a safe or an unsafe car is a subjective matter, and many state and national regulations are aimed at enforcing this requirement. The salesperson who sells an unsafe car might be sent to jail or given a fine if he or she violates this ethical objective. Hence, it is a good bet that the salespeople are familiar with the details of this area — not to say that all used cars are safe.

The used-car sales profession is replete with deceptive tricks, such as the bait-and-switch idea. Here, dealers advertise a splendid deal in the papers in order to attract buyers into dealerships. However, there are no comparable cars for the advertised price once the buyers arrive on the lot. Buyers are coaxed into looking at more expensive models for a look-see, and the hard sell begins. The public falls for these tricks over and over again. An ethical code of conduct would attempt to eliminate this behavior.

Other advice could be given about the repair of vehicles that are in acceptable working condition. You bring your car in for an ordinary tune-up and find out later at the office that a major brake job is needed. Do you trust the mechanic's word? What happens if you have taken the dealer's van to your office and there is no way to stop over and gain some physical evidence of the brakes' recent demise? Again, trust and truth telling are critical for ordinary business conduct.

So you have gone over these items one by one and put forth convincing cases that the used-car salespeople ought to behave in a "professional" and ethical manner. You state that all will benefit. The public will begin to trust the used-car salespeople and will begin to pay greater prices. There will be enhanced prestige for the salespeople and eventually greater salaries.

Unfortunately, the next day back on the lot, the salespeople must meet their weekly quotas. And it is a tough road to travel, knowing that they are professionals but also knowing that others who may not be so honest are getting many more sales. Eventually, our salespeople, who were so convinced by your speech, must pay their mortgages and find enough money to send their kids to college. And so they revert back to their old ways. These tactics may not be the most honest, but the salespeople are convinced that they are no more dishonest than the average used-car salesperson.

The purpose of this discussion is not to equate operations researchers with used-car salespeople. Operations research is generally practiced by extremely well-trained and well-meaning people, many of whom have Ph.D.'s from prestigious universities. These people are disciples of the scientific method and applied mathematics. Read the code of conduct that was developed by the Operations Research Society in the early 1970s.[8] Here, it is suggested that OR is an objective discipline in which the goal is to obtain "objective truths".

No, the purpose of my analogy is twofold. First, I would like to establish that improved professional conduct is unlikely to come from within, especially when the incentives are so strong in the opposing direction. The used-car profession is not likely to change itself without some powerful laws or threatening actions on behalf of the larger society. Why would the rascals change their behavior? Individuals are unlikely to cooperate in the prisoner's dilemma when there is no binding contract to act for the common good. Most people are myopic maximizers.

Second, I believe that the used-car analogy is most apt for describing the political process in the United States. The politicians should be admonished for not telling the truth and for being dishonest with the public — after all, we suffer by the low ethical conduct of our chosen representatives. How many Democrats and Republicans have quit for unethical conduct over the past years (Jim Wright, Richard Nixon, Spiro Agnew, to name a few prominent examples). These folks are the ones who need to improve their ethical conduct.

Unfortunately, many CDP's are employed by our politicians and become part of the political debate. Take the OTA and FORPLAN examples. It would be equivalent to having a used-car salesperson hire an operations researcher to "help" him or her show clients that all issues have been honestly and carefully considered. Eventually, the used-car salesperson will find an OR analyst (perhaps one who is down on his or her RFP luck) who will supply a CDP that produces the desired output. Again, the financial incentives strongly push the model away from the goals of scientific and objective analysis and towards the partisan ways of Washington politics. The OR model becomes a political weapon in this context.

This does not mean that all politicians and used-car salespeople are dishonest. There are many examples of exemplary conduct. (Recall Senator Proxmire's unbroken string of roll-call votes and his meager use of campaign funds — less than US $250 in one election). The reputations of these people speak for themselves.

It does mean that the use of CDP's in the public sector has the potential for causing harm by unethical and unprofessional conduct. The operations research field has barely touched on this critical issue. At some point, there will be a fatal error that will either precipitate grave consequences and outside pressure for regulation or will force the profession itself to recognize the possibilities and begin to develop strategies for reducing the potential for unethical conduct. Regardless, CDP's will need to be given greater scrutiny in the years to come.

NOTES AND REFERENCES

1. Wyscarver, R. A., *The Treasury Personal Individual Income Tax Simulation Model*, Secretary of the Treasury, U.S. Treasury Department, Washington, DC, 1984.
2. Mulvey, J. M. and Zenios, S. A., "Real-Time Operational Planning for the U.S. Air-Traffic System", *Applied Numerical Mathematics*, 3, 427–441, 1987.
3. Eiger, A., Jacobs, J. M., Chung, D. B. and Selsor, J. L., "The U.S. Army's Occupational Specialty Manpower Decision Support System", *Interfaces*, 18(1), Jan–Feb. 57–73, 1988.
4. Graettinger, J. S., "Results of the NRMP for 1980", *Journal of Medical Education*, 55(4), 382, 1980.
5. Mulvey, J. M. and Blount-White, S., "Computers in the Government: Modeling and Policy Design", *Public Productivity Review*, No. 42, 35–43, 1987, for details.
6. See Mulvey, J.M., "Strategies in Modeling: A Personnel Scheduling Example", *Interfaces*, 9(3), 66–77, May 1979, for more details about the evaluation process.
7. This approach is popular in finance. See Ingersoll, Jr. J. E., *Theory of Financial Decision Making*, Totowa, NJ: Rowman and Littlefield, 1987.
8. Operations Research Society of America, "Guidelines for the Practice of Operations Research", *Operations Research*, 19(5), 1123–1158, 1971.
9. Field, R. C., "National Forest Planning in Promoting U.S. Forest Service Acceptance of Operations Research", *Interfaces*, 14(5), 67–76, Sept–Oct 1984.
10. U.S. General Accounting Office, *Guidelines for Model Evaluation*, Washington, DC: U.S. Government Printing Office, 1979.

Chapter 3

How Do Values Become Incorporated in Models?

Part 1. Rhetoric and Rigor in Macroeconomic Models: Populist and Orthodox Swings in Latin America

Paul D. McNelis

INTRODUCTION

Many Latin American countries are in macroeconomic turmoil. Argentina, Brazil, Peru, and Nicaragua have gone through episodes bordering on hyperinflation. While Bolivia has successfully overcome hyperinflation, more than four years after its stabilization, stagnation continues, with yet no sign of positive per capita growth in income. Chile, while growing, only recently caught up with the per capita output losses it experienced in the last two decades. All of these countries must manage their economies with a high external debt burden.

Economic prescriptions for these countries range from populist solutions involving increased government activity (including nationalization of more enterprises) to traditional IMF prescriptions of tight monetary policy, reduced government activity, and promotion of export-oriented market-based activities.

In this chapter, I analyze the *rhetoric* embodied in the populist and orthodox international Monetary Fund (IMF) models that characterize (and polarize) policy discussion in Latin America, and I make an appeal for a new form of rhetoric as "disciplined conversation" for economic analysis and policy evaluation in Latin America. I describe the essentials of both the populist and the orthodox models of macroeconomic adjustment in the Latin American context. Then I analyze both of these models as forms of rhetoric or economic argument as persuasion. Finally, I provide an alternative model of macroeconomic adjustment, which highlights the role of income distribution as a macroeconomic variable. Until now, this variable has been left out of orthodox modeling altogether, while in populist proposals it serves as the centerpiece, but often in an *ad hoc* and unclear framework.

I contend that both the orthodox and the populist models are in need of ethical critique. The orthodox model ignores important variables, such as wage and income inequality, and thus sets the stage for increased social tensions. The populist model, on the other hand, highlights these effects while overlooking fundamental budgetary and external constraints, thus setting the stage for irresponsible policy actions. I propose that macroeconomic modeling, if it is to move beyond the deadlock of populist and orthodox swings, must explicitly build in and analyze the variables highlighted by populist models, such as wage and income inequality, while maintaining full recognition of the fundamental fiscal and external constraints in the macroeconomic system. I present preliminary simulation results of such a model in an analysis of proposals for debt relief in a model of a small, semi-open economy representative of Latin America.

In my analysis of the populist and orthodox alternatives, I call for rhetoric, defined as "disciplined conversation," in macroeconomic model development. Too often, populist and orthodox arguments are one-sided, and such conversation does not take place. This situation may be due to a false understanding of scientific methodology. In such a view, disagreements are due to either political differences, self-interest, or plain ignorance, since proper "objective methods" are, in this notion of science, sufficient to end all disputes about "facts". As I show in discussions of Latin American experiences with debt, stabilization, and stagnation, discussions of value, especially of equity and distribution, cannot be so easily separated from economic fact.

In this situation, parallel rhetoric, rather than rhetoric as disciplined conversation, may be the norm. Under parallel rhetoric, certain catch words or phrases used in arguments become so charged that they take on different meanings for opposing sides. Disciplined conversation as well as political solutions become extremely difficult, if not impossible. Charles Freund[1] first drew attention to parallel rhetoric as the cloud over the abortion debate that was preventing workable political compromises. As an alternative to the spread of parallel rhetoric, I argue for the rigor of disciplined conversation among macroeconomic model developers and policy analysts.

POPULIST AND ORTHODOX MODELS IN LATIN AMERICA

I begin with a discussion of the populist paradigm, particularly as practiced in Chile and Peru, and then turn to the International Monetary Fund (IMF) orthodox model, with an analysis of its overall track record and its Bolivian incarnation. Frequently, countries that followed the populist model were subjected to the orthodox models as a result of the economic and political instability at the end of the populist process. This order is often the sequential order for implementing these programs. For example, countries burdened by debt servicing may abandon their orthodox adjustment programs in favor of populism and debt moratoria.

Macroeconomic Populism

Dornbusch and Edwards[2] and Sachs[3] recently analyzed macroeconomic aspects of Latin American populism. Dornbusch and Edwards concentrated on the macroeconomic similarities of Salvador Allende's *Unidad Popular* (UP) program in Chile between 1970 and 1973 and Alan Garcia's *Alianza Popular Revolucionara Americana* (APRA) program in Peru since 1985. Edwards and Dornbusch admit that populism is a fuzzy concept. Citing Drake,[4] they isolate three elements: it uses "political mobilization, recurrent rhetoric and symbols designed to inspire the people"; it draws on a "heterogeneous coalition" aimed primarily at the working class and it connotes a "reformist set of policies tailored to promote development without explosive class conflict".[5]

Dornbusch and Edwards see the populist paradigm as a reaction against a monetarist experience of stabilization. They cite three ingredients: (1) initial conditions of slow growth, or outright depression; (2) a "no constraints" mentality, with idle capacity seen as providing a leeway for expansion, with room to squeeze profit margins through price controls; (3) policy prescriptions of reactivation through deficit spending, redistribution through wage and price controls, and restructuring of the economy, often through nationalization.[6] In their analysis of Chile and Peru, they identify four phases of the populist experience: (1) a period of vindication, with expansion and little inflation, as inventories and reserves are run down; (2) appearance of bottlenecks and shortages, an increase of inflation and a worsening of the budget deficit; (3) pervasive shortages, extreme acceleration of inflation, capital flight, and demonetization of the economy through the appearance of black markets with falling real wages, while political order disappears; (4) orthodox stabilization under a new government, with IMF assistance.[7]

What Dornbusch and Edwards did not mention in their analysis of

Chilean populism is that the Allende period was preceded by a Christian Democrat government that promoted many of the populist elements of the *Unidad Popular*. Brian Loveman writes:

> To a great extent the Christian Democratic program of political mobilization and deliberate "consciousness raising" (*concientizacion*) made impossible the attainment of its other major economic objectives such as control of inflation, increases in productivity, and higher levels of domestic savings and investment. With their hopes aroused by both the government's propaganda and the even more alluring Marxist vision in which redistribution of wealth and land would greatly improve the lot of the masses, Chilean workers and peasants could hardly be expected to accept government proposals for wage restraints, forced-savings plans, and moderation in labor disputes.[8]

In the presidential election campaign of 1970, the Christian Democratic candidate Radomiro Tomic promised further agrarian reform by expropriating all the rural estates "from the Andes to the sea", in what Loveman calls an attempt to "appear even more revolutionary than Allende".[9]

The Garcia APRA party was swept into office in 1985 after the economic decline of the *Acion Popular* program of Bellaunde from 1980 to 1985. Peru has perhaps the worst distribution of income in Latin America: the top one percent of the population takes roughly one half of the income. There was also an apathy among Peruvian economists to "orthodox" IMF programs implemented during the second half of the military dictatorship under Morales-Bermudez and in the Bellaunde administration. Dornbusch and Edwards point out that in 1982–1993, under an IMF program, real GDP fell by 16 percent and inflation almost doubled to 112 percent.[10]

The APRA program was spelled out in the *Plan Nacional de Desarrollo 1986–1990*. It emphasized full, participatory, and decentralized planning for stability, growth, distribution, and development; the primacy of the internal market over external markets; the necessity to redistribute income for sustained growth; and the necessity to reduce profit margins in order to generate higher capacity utilization.[11] The *Plan Nacional* denied the need for wage restraint, the inflationary effects of excess demand and fiscal deficits, the beneficial effects of increased interest rates on saving, and the positive effects of exchange rate depreciation on external accounts.[12]

What happened in Chile and Peru? In Chile by 1973, inflation reached 605 percent; growth fell to −5.6 percent; real wages fell by 30 percent relative to 1970; the budget deficit reached 25 percent of GDP; reserves fell to US $36 million, about 10 percent of the 1970 level; the trade balance went into a deficit of US $73 billion (from a surplus of $320 million in 1970); and the black market premium surged to over 2,000 percent. In Peru, inflation in 1989 was more than 5,000 percent, growth fell to −23.9 percent, and real wages fell about 50 percent below their 1980 average levels. Reserve levels actually became negative.

IMF Orthodoxy

The IMF policy prescription for countries experiencing inflation and a balance of payments deficit is a combination of fiscal/monetary austerity and devaluation. IMF policies usually recommend liberalization of internal, external, and financial markets to obtain further gains in output following the "adjustment program".

The adoption of an IMF program by a debtor country in return for credit comes in the form of a "stand-by agreement". The process by which the IMF provides financial assistance on the condition that client countries adopt particular policy actions is known as "conditionality". While the total credit provided by such agreements is relatively small, these "stand-by agreements" constitute a "seal of approval" for other public and private financial institutions to provide further credits to Latin American debtor countries. Thus, the IMF has considerable leverage over policy makers seeking credits, leverage that exceeds the credit it is actually able to provide, because IMF credits have "reassuring effects" on potential public and private lenders.

The IMF orthodoxy is a solution to the problem of matching policy targets with policy instruments. Typically, there are two policy targets: an external target (balance-of-payments equilibrium) and an internal target (control of inflation). There are two or more instruments: fiscal policy (control of the government budget deficit) and exchange rate policy (the implementation of a devaluation). Monetary policy is usually passive, adjusting to prior fiscal and exchange rate policies, as the money "printing press" continues to finance fiscal deficits. Reserves decline when there is a continued balance-of-payments deficit due to an overvalued currency or "bad" exchange rate policy. The decline in reserves puts downward pressure on the money stock. With continued high inflation, there is a demonetization of the economy. The stock of real money balances falls with rising prices, while domestic residents acquire foreign assets, gold, land, or other assets in place of money, as a hedge against inflation.

In the theory of macroeconomic policy, the number of instruments available to policy makers has to be greater than or equal to the number of targets. The IMF orthodox solution to the matching problem is to target the exchange rate policy to the external account and fiscal policy (meaning fiscal stringency) to the control of inflation.[13]

The IMF orthodoxy is a program of austerity. The devaluation reduces the purchasing power of wages, as prices of tradeable goods rise, and thus cuts the standard of living of domestic residents.

Along with these macroeconomic policy reforms, IMF programs recommend liberalization of internal and external markets. This involves decontrol of administered prices and wages, the removal of credit rationing in favor of flexible interest rates, dismantling of a quota system in favor of a uniform tariff, and later a reduction of this tariff towards a freer trade system. IMF programs also advocate the abolition of exchange controls and the opening of the capital market. All of these structural reforms are designed to make the allocation of resources more "efficient", and thus increase output and productivity, which in turn contribute an increase in supply and a lower rate of inflation. IMF orthodoxy is thus inconsistent with any form of price or wage controls or with other forms of government intervention designed to break inflationary expectations at the start of a stabilization program.

If the stabilization and liberalization reforms of the IMF program are followed at time t^*, then the IMF "model" promises a higher level of output at time $t^* + 2$ (relative to time t^*), after a brief fall in output at time $t^* + 1$, as well as lower inflation. If policy makers only adopt the macroeconomic reforms without the structural reforms, output at time $t^* + 2$ will be equal to its value at time t^*, and inflation will be permanently lower. Proponents of the IMF model envision a two- to three-year time horizon from time t^* to $t^* + 2$, so that the recession should not last more than one year or so.

Dornbusch[14] and Edwards[15] have presented critiques of the IMF models and policies, based on analyses of papers written by IMF researchers and interviews with IMF staff members in operations.

Whatever problems one may have with IMF models or methods, the track record since the debt crisis of 1982 has not been impressive. For example, the degree of fiscal policy "compliance" with conditionality for the 34 approved programs was only 30 percent in 1983, 19 percent in 1984, and 43 percent in 1985. Not surprisingly, the percentage of inflation targets that were met or exceeded for the same programs was only 48 percent in 1983, 41 percent in 1984, and 36 percent in 1985.[16] While the populist macroeconomic programs exemplified by Allende and Garcia appear to be recipes for disaster, the alternative program of IMF orthodoxy does not give much assurance of success.

The IMF model is open to attack on several points. While few, if any, economists (outside the proponents of populist programs) deny the necessity of fiscal reform for inflation stabilization, the usefulness of devaluations is open to debate. In countries where a large number of imported inputs (such as steel, electronic components, or energy) are needed for domestic production, a devaluation will raise the costs of production and thus lead to production cutbacks and more unemployment, at least in the short term. Furthermore, liberalization of financial markets and removal of exchange controls can lead to abrupt capital outflows (if there is sufficient doubt about the resolve of the policy makers) and thus loss of resources needed to finance domestic investment. Alternatively, liberalization can precipitate abrupt capital inflows (if expectations are good), which lead to an appreciation of the real exchange rate (the relative price of tradeable goods to non-tradeable goods) as the inflows generate demand in the service and non-tradeable sectors of the economy. This change in the real exchange rate (or simply the ratio of traded to nontraded goods prices) will generate a decline in investment and production in the tradeable-goods sector, with still more unemployment, since the "tradeable-goods" sector is the labor-intensive manufacturing sector of the economy. It is thus important for policy makers to avoid this type of real exchange rate appreciation, since it will only increase the employment costs of reducing inflation.[17] Similarly, the abrupt freeing of domestic interest rates from controls may lead to "upward financial repression" after a long period of controls and credit rationing. Interest rates overshoot during the "learning period" when financial markets adjust to a new climate of competition. For this reason, contrary to IMF orthodox teaching, many economists now believe that the financial sector and exchange market should be the last sectors to be liberalized.

The New Economic Policy program of the Victor Paz Estessoro *Movimento National Revolucionario* government of 1985 reduced inflation from a high of 34,000 percent to less than 20 percent per annum by 1989, one of the lowest rates in Latin America. The Bolivian debt crisis occurred in 1979, well before the Mexican crisis of 1982, when banks suspended new credits to state enterprises. Populist governments responded by supporting government enterprises with domestic monetary expansion in the early 1980s, and the inflation rate accelerated, as the tax base shrank and capital flight mounted.

While the Bolivian stabilization was not an official IMF program, the policy package contained in *Decreto Supremo* 21060 of the Paz Estessoro government embodied many of the orthodox IMF recommendations. The government cut the deficit, removed or reduced administrative price controls, made labor markets more "flexible" (mainly through the massive layoffs of miners in COMIBOL, the state-owned mining corporation, which fell from a workforce of over 30,000 to less than 7,000, and by the removal of mandatory worker "tenure" and mandatory bonuses in all private-sector contracts), and completely opened exchange and financial markets.[18] The government also cut tariff rates to a uniform 20 percent on all items and removed quotas and licensing requirements.

While the inflation record has been impressive, growth has still not returned, more than four years after the stabilization. Output continued to decline in per capita terms until 1988, and from 1988 to 1989 output growth just matched population growth. Since 1980, public consumption (which includes the provision of social services in education, nutrition, health, and sanitation) has fallen by 17 percent, and public investment (which includes investment in transportation, communication and infrastructure) has fallen by 61 percent.[19]

One of the expected benefits of financial liberalization is the convergence of domestic interest rates to international rates of interest, adjusted for the rate of devaluation. Financial sector openness, in theory, should make the domestic financial markets more competitive and efficient and thus drive down interest rates to the levels given by international arbitrage. Figure 1 shows that this convergence has not occurred in Bolivia. The LIBOR rate (representing the foreign rate of interest) plus devaluation was about 9 percent

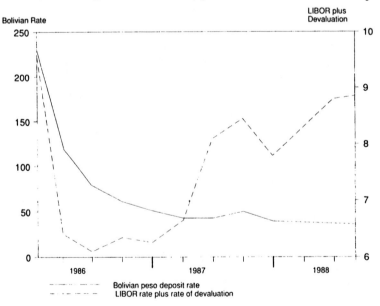

Fig. 1.

at the beginning of 1989, while the domestic real rate of interest was close to 30 percent. Bolivia thus suffers from "upward financial repression".

As for the behavior of the real exchange rate, Morales[20] has argued that the *peso boliviano* is overvalued by about 10 percent, and has had sustained recessionary effects on production and employment.[21] He does acknowledge the trade-off between export competitiveness, reactivation and growth policies (involving a real exchange rate devaluation), and continued price stability. The key problem facing Bolivian policy makers is how to resolve this conflict and to administer the exchange rate in this context.

While Bolivia may appear to be an orthodox showcase model for inflation stabilization, as Dornbusch has pointed out, the stark fact is that Bolivia has not fallen back on its feet running, but has fallen into a hole. This does not mean that inflation stabilization is wrong — macroeconomic instability is never a good thing — it does raise the question of finding other policies that can promote growth and guarantee continued stability.

POPULIST AND ORTHODOX MODELS AND RHETORIC

The failure of both the populist and IMF orthodox models to bring both stability and growth gives some credibility to the arguments of McCloskey[23,24] that economists must recognize the place of rhetoric and persuasion in economic argument.

McCloskey argues that in their actual scientific work economists argue about the "aptness of economic metaphors, the relevance of historical precedents, the persuasiveness of introspections, the power of authority, the charm of symmetry, the claims of morality".[25] He believes that nothing is gained from clinging to the Scientific Method, or any methodology, except that of "honesty, clarity, and tolerance", through rhetoric, defined as "disciplined conversation" or "exploring thought by conversation".[26] Economists should not discriminate against propositions on the basis of "epistemological origin", since not all regression analyses are more persuasive than moral arguments, nor controlled experiments more so than introspections; some subjective, soft, vague propositions are more persuasive than some objective, hard, precise propositions.[27] McCloskey believes that economic argument should be examined from this humanistic side. We can still see the syllogisms, but also the choosing of metaphors, the telling of stories, and the appealing to character, through this "richer way" of seeing economic analysis.[28] He reacts against any philosophy of science that pretends to "legislate the knowable", since the claims of an "overblown methodology of Science merely end conversation", and non-argument becomes the necessary alternative to narrow argument under this standard. He maintains that his cure would not dispose of the "illuminating regression, the crucial experiment, the unexpected implication unexpectedly falsified", since these too persuade scholars.[29]

McCloskey's central thesis is that "rule-bound" methodologies in economics are unreasonable, that the "persuasiveness-independent notion of truth" should be rejected.[30] He identifies two major defects of this "modernist" rule-bound approach to scientific methodology: the impossibility of falsification and the superficiality of prediction as a scientific aim.[31]

Rappaport argues that the "epistemological approach" is broader than the "modernist" or "rule-bound" approaches that McCloskey criticizes, and that it has the advantage of preserving the connection between economics and the pursuit of truth, which McCloskey sets loose.[32] To this contention, McCloskey replies that there is no "Archimedean point" outside conversation from which to level up the world.[33]

Rosenberg challenges McCloskey on the role of predictions:

No matter how many sorts of knowledge there may be—the kind for which we look to economics is the kind controlled and certified by a reality that neoclassical economics did not create but found pre-existent. Improvements in knowledge of this reality can only derive from predictions, successful and unsuccessful.[34]

McCloskey replies to Rosenberg's criticism in the form of a dialogue between Socrates and Ion:

SOCRATES: Science is defined to be what is predictive, in the pattern of the 19th century's understanding of certain branches of physics. So when science proves to be nonpredictive or in some other way deviant from this model, it is the scientist, not the philosophy of science, that is defective.

ION: Certainly. At last you are hearing the point.

SOCRATES: In other words, you can second-guess the artists and scientists of the world. That is truly wonderful.

ION: Yes, so it is.

SOCRATES: I admire your wisdom more than I can say. Yet there is one respect in which I blame you: in our conversation you have quite unjustly neglected to reveal to me the method of achieving this profitable wisdom in second guessing. ...I conclude that you are either a man unjust or a man divine. What it is to be?

ION: It is far lovelier to be deemed divine.

SOCRATES: Then this lovelier title, Ion, shall be yours, to be in our minds divine, and not a scientist, in praising science.[35]

McCloskey argues for a form of rhetoric disciplined by moral philosophy in economic analysis:

The classical problem was that rhetoric was a powerful device easily misused for evil ends, the atomic power of the classical world, and like it the subject of worry about its proliferation. The solution was to insist that the orator be good as well as clever: Cato defined him as "*vir bonus dicendi peritus*", the good man skilled at speaking, a Ciceronian ideal as well. Quintilian, a century and a half after Cicero, said that "he who would be an orator must not only *appear* to be a good man, but cannot *be* an orator unless he *is* a good man" (*Institutio* XII, 1, 3).[36]

McCloskey notes that the canonized methods of science can be and have been methods of deceit, and the Greeks and Romans were more sensitive to these possibilities and "less hypnotized by the claims of method to moral neutrality".[37]

Whether or not "theoretical realism" is compatible with the rhetorical approach advocated by McCloskey is explored by Maki.[38] In the McCloskey framework, theory is "world-making" or instrumentalist rather than "world-discovering", since there is no "Archimedean point" from which to evaluate the "real world" apart from the conversation of economists. Maki believes that "marketing rhetoric with explicit theoretical anti-realism" is not good rhetoric.[39] Maki thus supports realism with rhetoric in economic analysis.

I think that the populist and orthodox models of macroeconomic policy in Latin America illustrate well many of the problems that McCloskey wishes to address with his rhetorical analysis. I also do not think it matters much whether one is a theoretical realist or an antirealist. Questions of poverty and inequality in Latin American involve explicit value judgments both in the formulation of policy and in the design of proper indicators for measurement.[40] Furthermore, these questions are at the center of academic discussion, political debate, and radical mobilization. Thus, as pointed out by Maki, one can be, like McCloskey, a "non-Cartesian" about the cognitive behavior of economists while being Cartesian about the theories propounded by economists.[41] Poverty and income distribution questions would still be important. If one were non-Cartesian about both, the questions would be even more important.

The IMF model, which emphasizes budget constraints, external sector identities, and the gains from promarket liberalization policies (without explicit dynamics), ignores questions of "value", such as income distribution and equity, in favor of tractability and general applicability to a wide variety of countries and circumstances. Thus, IMF teams concentrate on variables fairly easy to measure, such as the money supply, price indices, and the exchange rate, and they use "financial programming" to examine the interaction of these variables with general accounting identities and budget constraints for the public sector. By contrast, the populist model, in reacting against the IMF orthodoxy, emphasizes variables less tractable and thus less easy to measure (such as income distribution and poverty), and stresses particular circumstances of the economy that make their case either "special" or an exception to the general diagnosis and treatment given by IMF "financial programming".

The rhetorical approach, whether realist or Cartesian, helps us to see problems in orthodoxy and populism. The IMF approach would view the populist model as "unscientific" and thus would find populist critiques inadmissible, while populists would consider the IMF approach basically a conservative economic and political ideology with scientific window-dressing.[42] There certainly is "rhetoric" in both the populist and IMF models, which may be either evidential (such as logical rigor in IMF models) or non-evidential devices (such as irony or sarcasm in populist attacks against orthodox models).[43] There is parallel rhetoric, but little, if any, "disciplined conversation", among proponents of these approaches when it comes to evaluating IMF performance or pitfalls in populism. Words such as "feasibility" and "distribution" have become so charged that a minimal common vocabulary or glossary of terms is elusive or non-existent. A shift from parallel rhetoric to rhetoric as disciplined conversation would go a long way towards the development of newer, richer models, with better chances of success, whether we view these models as "world transforming" or "world discovering" or both.

AN ALTERNATIVE MACROECONOMIC FRAMEWORK FOR LATIN AMERICA

I propose an alternative macroeconomic framework for analyzing adjustment in Latin America, one which combines the fundamental identities as well as budget and supply constraints of IMF orthodoxy (and thus feasibility conditions) with the concern of populist models for income distribution as a macroeconomic issue.

To be sure, macroeconomic analysts have not ignored income distribution and inequality. Berg and Sachs,[44] for example, have cited income inequality, measured by the ratio of income shares of the highest quintal to the lowest quintals in the population, as a statistically significant determinant of rescheduling of international debts in a cross-sectional study of countries with international debts. Similarly, Bourguinon, Branson, and de Melo[45] examined income distribution and adjustment following stabilization programs, using a simulation model. They concluded that if these programs do not contain specific components targeted towards the poor, there will be some form of "permanent damage" for those below the poverty line.[46] These contributions, however, are quite minor in comparison with the vast literature that has by and large overlooked income distribution and inequality as macroeconomic issues.

To illustrate the importance of inequality and income distribution, I introduce this variable into the analysis of debt/equity swaps, a program for reducing the debt burden on Latin American countries through debt conversion.

Modeling Inequality and Debt/Equity Swaps

Extending previous joint work,[47] I incorporate inequality into macroeconomic analysis of debt/equity swaps. This type of swap is an exchange of outstanding debt of a Latin American country held by a commercial bank in the secondary market for cash with an investment firm. The investment firm, in turn, has the right to redeem the face value of this debt, or a reduced fraction thereof, for domestic currency of the debtor country for the purchase of shares in domestic industry. There are usually some restrictions on the equity purchases, such as a moratorium on the repatriation of profits for a period of 10 years following the equity acquisition.

While the debt/equity swaps have not been large relative to the total amount of Latin American debt, such swaps have been large in Chile, which has reduced about 40 percent of its debt through such operations. Brazil has also been active in this type of debt conversion.

Much of the analysis of debt/equity swaps has concentrated on the behavior of asset prices in the domestic economy following these actions or on the distribution of "welfare gains" between debtor and creditor countries before and after the swap.[48] While there has been some populist criticism, centering on the transfer of ownership of domestic resources of debtor countries to foreigners, there has been little analysis of the income distribution effects of these operations within the debtor countries, either by orthodox or populist modelers. Thus, analysis of this issue may be a good beginning for a "disciplined conversation" between populism and orthodoxy.

Without belaboring the details of the model, I note that the identities, budget constraints, and supply functions emphasized by orthodox macroeconomic models appear in the model. However, income distribution and inequality appear as well, both as variables that respond to policies and as variables that have feedback effects on the rest of the system, particularly on output supply.

Income distribution and inequality respond to two factors: inequality in wages across sectors and inequality in the distribution of profits. In this model, I incorporate both of these factors. In McNelis and Nickelsburg,[49] only wage inequality affected overall income inequality.

There are $(m + 1)$ unions, so the wage contour is a vector of $(m + 1)$ nominal wages, assumed to be staggered over $(m + 1)$ periods. I assume that the membership of the labor force is equally distributed over the $(m + 1)$ unions. To simplify the simulation exercises still further, there are 800 workers evenly divided among 8 unions, and there are an additional 100 whose income is from their earnings on capital returns. This contour is given in equation (1), below. The individual wage negotiated at the present, $W(0,t)$ is fully indexed to the price level, and also reacts to differences between actual output supply and full-employment (or full capacity) output, y^f. This behavior is illustrated by equation (2). The average wage is simply the mean nominal wage. The income of the worker is the wage income plus returns from the share of the particular working sector in the ownership of capital. This identity is shown in equation (3). The income for the capitalist sector is simply the share of this sector in the returns on capital, given by equation (5). Returns to capital are simply output less the wage bill less the current account and less whatever claims are given to foreigners in debt/equity swaps. This is shown in equation (4). The average income is simply the income given to each group multiplied by (1/9), since each group represents this fraction of the economically active population. Equation (6) shows this definition. The coefficient of variation of income is simply the standard deviation divided by the mean. This index of inequality appears in equation (7).

The following equations thus describe the way inequality is determined in the macroeconomic system:

Wage contour:

$$[W(m, t), W(m-1, t), ..., W(1, t), W(0, t)] \tag{1}$$

where $W(i, t)$ is the wage of union i at time t.

Individual wage adjustment:

$$W(0, t) = P_{t-1} - \frac{\alpha_w}{(y_{t-1}^5 - y_{t-1}^f)} - \alpha_0 \tag{2}$$

where p is the price level, y^s is the output supply and y^f is the long-term full employment.

Income to each working-class sector:

$$Y(i, t) = w(i, t) + \sigma_i\, YCAP \tag{3}$$

where $YCAP$ is capitalist income, and σ^i is the share in capitalist income of sector i.

Total capital returns:

$$YCAP = Y - \sum W(i, t)\, L_i - B - EXD_{t*}\, \alpha^* (1 - \lambda)\, \lambda^{t*-t}$$

$$\text{where } EXD_{t*} = \begin{cases} EXD_t & \text{following debt/equity swap} \\ 0 & \text{otherwise} \end{cases} \tag{4}$$

with EXD representing the outstanding stock of debt of the country, α^* the percentage swapped, B the current account balance, L_i the number of workers in sector i, and λ the rate at which the foreign purchase of equities is phased into domestic ownership of equities.

Income to capitalist sector:

$$Y_c = \sigma_p\, YCAP \tag{5}$$

where σ_p is the share of capitalists in the total returns to capital.

Average income:

$$\overline{Y}_t = [\sum_{i=0}^{m} Y(i, t) + y_c]/(m + 2) \tag{6}$$

Coefficient of variation of income:

$$CV_{y,t} = \{\sum_{i=0}^{m} [Y(i,t) - \overline{Y}_t]^2 + [Y_c - \overline{Y}]^2/\overline{Y}_t^2 \cdot (m + 2)\}^{-\frac{1}{2}} \tag{7}$$

I measure inequality by the coefficient of variation, rather than the Gini coefficient, calculated from the Lorenz curve, for the sake of simplicity. I also use the ratio of total labor income to total output as an index of distribution or social tension.

The coefficient of variation can change from wage inequality or from changes in the distribution of returns to capital among workers and capitalists. Wage inequality can increase from sudden changes in inflation rates due to backward-looking indexation. A sudden change in inflation reduces the real wage of certain sectors locked into longer-term contracts before the inflationary surge took place while giving a chance for the sector currently negotiating its wage to move forward. A debt/equity swap, which transfers ownership to foreigners, will change the profile of returns to capital among workers who own shares and among capitalists whose only income consists of their returns from capital. Since a debt/equity swap may also have inflationary consequences, and thus increase wage inequality, this action will affect inequality through both of these channels.[50]

I also assume that income inequality has effects on aggregate supply or production. Current period output depends negatively on the real wage and the real exchange rate (which increase costs of labor and imported inputs) and on the coefficient of variation of income. Increases in inequality will increase social tension and increase work stoppages through strikes and prolonged contract negotiations. Finally, capital inflows, which increase external indebtedness, may also have some positive effects on output through their effects on productivity, if the borrowing is used to finance infrastructure investment. I assume that there is some productivity effect, but that this effect declines exponentially.

Production thus depends on costs, both internal due to wages and external due to exchange rates, as well as to inequality and to capital inflows. The following equation thus describes supply at the current period:

$$y = \alpha_0 + \alpha_1 (W/P)_{t-1} + \alpha_2 (EXR/P)_{t-1} + \alpha_3 CV_{y,t-1} + \alpha_4 \sum_{i=1}^{n} \beta^i EXD_{t-1} \qquad (8)$$

where $\alpha_1 < 0$, $i = 1, 2, 3$, and $\alpha_4 \geq 0$.

Dynamic Simulation: Before the Crisis

I set up the crisis situation with debt servicing by simulating the model from an initial stationary equilibrium in which borrowing from abroad begins and continues at a rate of 2 percent of gross national income per year. This borrowing or capital inflow continues until net transfers become negative. Net transfers are defined as capital inflows less the interest on outstanding debt. Initially, these transfers are positive, as the inflows dominate the initial interest payments. However, as the stock of debt accumulates, the transfers fall in value and eventually become negative, as the interest payments on the accumulated debt eventually dominate the value of the continuing inflows.

Figures 2 and 3 picture the adjustment of the price level, the real wage, and the real exchange rate, as well as the two income distribution indices: the coefficient of variation and the share of labor in total income.

These figures show an increase in the price level and the real wage, and a fall in the real exchange rate, indicating an appreciation in the real exchange rate with increased domestic purchasing power over foreign-produced imported goods. This represents an increased standard of living as a result of the increased borrowing from abroad, but also some inflationary pressure.

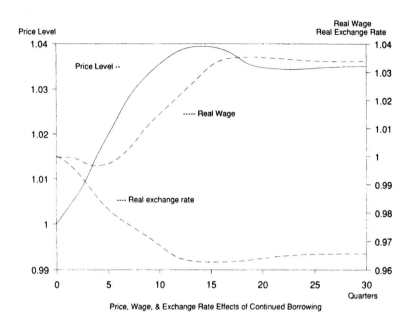

Price, Wage, & Exchange Rate Effects of Continued Borrowing

Fig. 2.

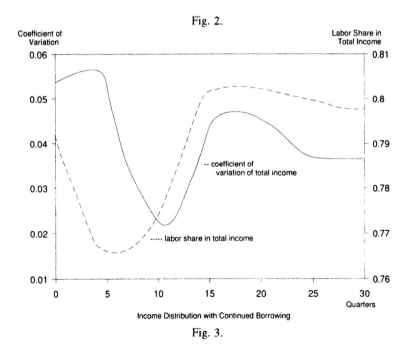

Income Distribution with Continued Borrowing

Fig. 3.

This combination of currency appreciation, increased standard of living, and inflationary pressure characterizes well the experiences of the "Southern Cone" countries (Argentina, Chile, and Uruguay) in the late 1970s and the early 1980s, when controls on foreign capital inflows were removed and massive new lending took place from the deregulated eurodollar banks. The incentives for the Latin American countries to borrow were particularly strong, since the interest rates were particularly low.

For the income distribution indices, there is cyclical behavior in both the coefficient of variation and in the labor share or total income. Inequality increases and the labor share falls as borrowing increases productivity and the return to capital. However, inequality falls and soon the labor share increases, as the real wages increase. The longer-term effect before the crisis is a slightly higher labor share in total income and a slightly lower index of inequality.

Sooner or later, however, the net transfers will fall and become negative unless the creditors are willing to lend larger and larger amounts. In 1982, the commercial banks suspended new lending, and net transfers became negative for the first time.

In this simulation model, I stop the dynamic adjustment when the transfers become negative and implement a policy reaction, in this case a debt/equity stop. Of course, the debtor country can continue to service the large debt and undergo a period of austerity in order to earn greater export revenue, or the country can suspend debt servicing and face the possibility of creditor country sanctions.

The debt/equity swap in this simulation run consists of a transfer of ownership of one-half of the existing stock of debt to foreigners, while new borrowing is terminated. Thus, the negative net transfers are reduced but not eliminated, so that austerity will characterize the adjustment process.

Dynamic Simulation: After the Crisis

Figures 4 and 5 illustrate the adjustment of the price level, real wage, real exchange rate, and income distribution indices after the debt/equity swap.

At the time of the debt swap, all variables shift to new "initial conditions". In both Figures 4 and 5, time "0" now indicates the new starting point for adjustment after the suspension of new borrowing and the debt/equity swap. It is not surprising that the price level and real wage fall and the real exchange rate rises. Since capital inflows have stopped, deflationary pressure exists on prices and wages, and no forces lead to an appreciation of the domestic currency.

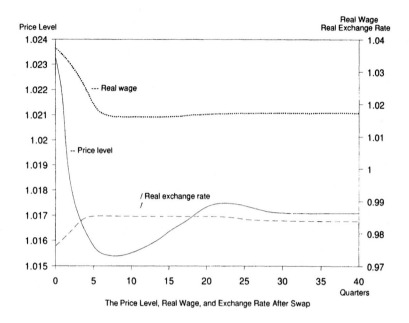

The Price Level, Real Wage, and Exchange Rate After Swap

Fig. 4.

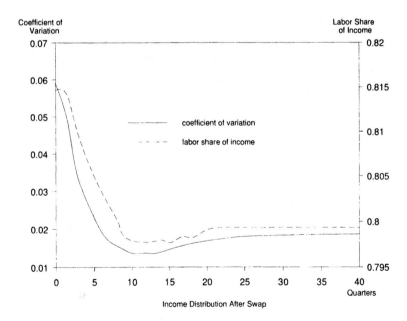

Income Distribution After Swap

Fig. 5.

The real wage falls in a monotonic path while the real exchange rate rises and the price level falls below its longer term value. A deflation thus occurs, along with a cut in the standard of living through lower real wages and a "weaker currency" with less purchasing power over foreign goods.

As for the adjustment of the income distribution indices, they also experience discrete changes at the time of the swap. At the new initial condition, time "0" in Figures 4 and 5, the coefficient of variation and the labor share are slightly higher than their corresponding values at the end of borrowing period. The jump in both variables is due to the cut in capital returns to domestic residents, since a share of income is now transferred to foreigners through the swap. Thus, the labor share of domestic income is now larger. The coefficient of variation rises, since the change in the price level affects the profile of real wages distributed across unions, with unions locked into longer-term contracts experiencing an increase in real earnings and unions currently negotiating contracts losing such an increase due to indexation agreements that tie wages to price changes, both positively and negatively.[51]

After the debt/equity swap, both variables fall. Income distribution becomes more equal, as measured by the coefficient of variation, but the labor share in total income also falls. This situation is more egalitarian, but the total worker share is lower.

If "workers" are poor, or if those who work and receive wages represent the poor while those who live solely on returns from capital are considered "rich", then the results indicate that the poor have relatively lower claims or entitlement to national income. The net effect of the suspension of new borrowing and the debt swap is to redistribute income away from the poor to the rich. By any welfare criteria, based on relative deprivation indices, this adjustment is socially and politically suboptimal. Thus, a case exists for some type of government redistributive policy, through taxation and transfers, in order to ease the relative debt burden placed on the "poor". A case also exists for forgiveness of a part of the debt to ease the burden of adjustment.

Rhetorical Elaboration of Dynamic Simulation Results

The results given by the dynamic simulations represent the outcomes of a disciplined synthesis between populist concern with inequality and orthodox emphasis on fundamental supply constraints and accounting identities.

The figures show that income distribution, measured by the coefficient of variation and labor share in national income, follows a cyclical adjustment and thus has a dynamic pattern that has so far attracted little interest in macroeconomic analysis.

Adjustment of the price level, real wage, and real exchange rate variables show that there is indeed austerity after the crisis, just as the borrowing brought in a period of relative prosperity, indicated by higher real wage and real exchange rates combined with lower longer-term inequality and a higher labor share of national income. The basic story the model tells is that this cannot last forever, unless creditor nations are willing to lend in larger and larger amounts. When the crisis comes, both the overall standard of living and income distribution indices in the debtor nations will change. An abrupt change can easily provoke social tension, given the expectations of rising living standards, and lead to strikes, class conflict, and political populism, as it did in Bolivia and later in Peru, with macroeconomic instability and hyperinflation. Political repression may go hand in hand with fundamental economic reform following the period of chaos, as it did in Chile.

Adjustment to the debt crisis is hardly a mechanical process easily predicted by a deterministic numerically specified model. With deteriorating conditions, populist policies, hyperinflation, and finally orthodoxy may form part of the social–political–economic cycle. The simulation model thus functions as an indicator of early warning signals.

The results illustrate the need for further disciplined conversation on the design of social pact, that can reduce the tensions and distributive conflicts that are on the horizon. As Sachs[52] has pointed out, these social pacts and redistributive institutions, so successful in moderating conflicts in the Scandinavian countries, have been markedly absent in Latin America. This absence may explain the recurring populist/orthodox cycle in Latin American macroeconomic management, across countries and across decades. It may also explain why the Israel stabilization plan of 1985 worked, while similar plans for Argentina and Brazil in 1985 and 1986 did not.[53]

The modeling of this section is thus a form of rhetoric, and an argument for political action, both on regional and international levels. If the populist/orthodox cycle is to be broken, then there is need to moderate the conflicts bound to occur when standard-of-living and income-distribution indices deteriorate. The current debt crisis has triggered such a situation, which persists even when debt/equity swaps occur.

The scope for alleviation of poverty and inequality by governments pressed by debt servicing payments and low tax revenues is small. Thus, creative and cost-effective action is needed in Latin America, especially in the areas of health, education, sanitation, and nutrition, which are the first to deteriorate as the standard of living declines during the adjustment process.

Fortunately, a model and an infrastructure is in place in Latin America that does deliver these services, at low cost, to the poor and least advantaged. This is the network of Basic Communities, supported by the Catholic Church and staffed by religious and pastoral agents who live with the poor on minimal support. While I am sure that pastoral agents cherish their independence from civil authority and do not wish to be incorporated into a state program, their experience, infrastructure, and delivery systems of important social services are worth studying, learning from, and even imitating, *mutatis mutandis,* in the design of social programs targeted at the plight of the poor. The Catholic Church thus functions in Latin America as a redistribution institution, and its experience may be beneficial for the design of more, in political discussion and social pacts among conflicting elements.

This may mean, for Latin American countries, a social pact entailing social service, much like a domestic Peace Corps, for young men and women who have had the benefits of a higher or technical education and who can provide the needed services of education, basic health care, sanitation, and nutrition. A concerted effort of social service may be an important instrument of redistributive conflict resolution in Latin America and may break repeated cycles of populism and orthodoxy in macroeconomic policy. It may also pave the way for longer-term economic growth in Latin America.

It is well known in development theory that investment in people, in the human resources of the economy, is the key to long-term growth as well as social stability. The United States was the beneficiary of a large wave of skilled and semi-skilled immigrants in the nineteenth century. These immigrants built the railroads, dug the canals, mined the coal, grew the crops, and sent their children to land-grant colleges. For Latin America, it is the poor, the people at the periphery, who will have to be brought into full participation in national life and economic reconstruction. This means that their traditions, values, and cultures will become a more significant part of development as their energies are channeled into the economic growth process.

There may be other social pacts for redistributive conflict resolution and other paths to growth and development. However, given the severe budgetary conditions and the limited scope for new international borrowing, Latin American nations may have to look to their own human resources and direct their efforts there, in cost-effective ways, through social service programs modeled on the basic communities of the Catholic Church, to find some measure of stability, growth, and hope in the coming decades.

Table 1. Dynamic Profiles of Borrowing, Orthodoxy and Populism in Latin America

Social Indicator	Proxy Variable	Pre-Crisis I Borrowing Phase	Post-Crisis I Orthodox Phase	Post-Crisis II Populist Phase
Standard of living	Real wage Real exchange rate	Rising standard of living through rising real wage appreciating exchange rate	Falling real wages and depreciating exchange rate imply falling living standards	Policies initially aimed at raising real wages and improving living standards with neglect of reserves
Macroeconomic instability	Inflation	Rising but accepted inflation	Accelerating inflation rates induce stabilization program	Initially constant, but sharply accelerates
Political viability	Parallel rhetoric Disciplined rhetoric	Low level of debate Consensus	High level on few issues	Intense level spread over many issues Loss of rigor
Labor tensions	Income inequality Labor share of income	Falling indices Rising share	Fluctuating index Falling share	Initial favorable change but soon reversed
Redistributive institutions	Formation of social pacts	Little interest or concern	Increasing attention and demands made on organizations	Roles are replaced by direct government actions Attempt to co-opt independent services

Dynamic Profiles of Borrowing, Orthodoxy and Populism: The Role of Rigor

Table 1 summarizes the broad features of my analysis of borrowing, orthodoxy, and populism in Latin America. On the vertical axis, I plot selected social indicators or roles for public institutions, while on the horizontal axis I plot three policy phases: precrisis borrowing, postcrisis orthodoxy, and postcrisis populism.

The social indicators fall into five broad categories: standard of living (as given by real wages and the real exchange rate), macroeconomic instability (measured by inflation), social tension and conflict (measured by income inequality and the share of labor in national income), political viability (measured by two forms of argument: parallel rhetoric and disciplined conversation), and social resiliency (measured by the functioning of redistributive institutions).

Table 1 shows that when the standard of living is high and macroeconomic instability and social tension are low, people have little interest in redistributive institutions. Parallel rhetoric is unimportant, and with strong consensus, there are few incentives for disciplined conversation. In the postcrisis orthodox policy phase, there is an initial situation of falling living standards, a strong focus on macroeconomic instability through anti-inflation policies, and increased social tension as the labor share of output falls. Parallel rhetoric increases as special interests form pressure groups, and disciplined conversation diminishes. In this phase, the role of redistributive institutions, both in moderating conflict and in providing a safety net, becomes important. If these institutions do not function effectively, there will be another phase: postcrisis populism. In this phase, the standard of living will temporarily improve due to government-mandated wage increases, with little change in the short term in inflation and macroeconomic instability. Parallel rhetoric will rise to intense levels as previous orthodox approaches are ridiculed, making disciplined conversation unlikely. The role of redistributive institutions will be co-opted, and possibly weakened, by government action.

What Table 1 illustrates is the need for rigor, which supports and strengthens the rhetoric of disciplined conversation, in macroeconomic modeling and development. In periods of rising living standards, little attention is paid to the rigorous disciplined conversation of social and political sciences or to the benefits of stable redistributive institutions. However, it is precisely at this time that there is a need for reinforcement, because their success or failure in the following orthodox phase is crucial. If they do not do the job of moderating parallel rhetoric and social tension, the populist reaction may be inevitable. This reaction, in turn, will further weaken disciplined conversation and the resilience of redistributive institutions.

This orthodox–populist swing in policy programs is of course a recurring phenomenon in Latin America. My analysis shows the need for rigor in rhetoric (through disciplined conversation) as well as for the development of redistributive institutions.

CONCLUSION

By orthodox standards, my final policy proposals for new redistributive institutions and social pacts in Latin America may have strayed well beyond the scope of proper scientific rigor, even if it is based on the rigor of disciplined conversation. By the same token, populist sentiment may not welcome my acceptance of fundamental budget constraints and identities emphasized by IMF teams when negotiating "stand-by agreements". So be it. My point is that not all sound economics lead ethical policy choices, but that ethical policy cannot be inconsistent with sound economics. My hope is that this analysis will encourage broader and more rigorous disciplined conversation among social scientists and policy analysts, while taking into account the experiences of the Catholic Church (as a redistributive institution) and the changing forms of rhetoric as important social indicators in the macroeconomic adjustment and growth process.

NOTES AND REFERENCES

1. Freund, C., "Parallel Lines of Debate on Abortion", *The Washington Post*, April 25, 1989.
2. Dornbusch, R. and Edwards, S., "Macroeconomic Populism in Latin America". Unpublished manuscript, 1989.
3. Sachs, J., "Social Conflict and Populist Policies in Latin America". *National Bureau of Economic Research Working Paper No. 2897*, 1989.
4. Drake, P., "Conclusion: Requiem for Populism," in M. L. Conniff, ed., *Latin American Populism in Comparative Perspective*, Albuquerque: University of New Mexico Press, 1982.
5. Dornbusch, R. and Edwards, S., *op. cit.*, 3.
6. *Ibid.*, 5–6.
7. *Ibid.*, 6–7.
8. Loveman, B., *Chile: The Legacy of Hispanic Capitalism*, Oxford: Oxford University Press, 285–286, 1988.
9. *Ibid.*, 293.

10. Dornbusch, R. and Edwards, S., *op. cit.*,. I have written about the problem of "dollarization" in Peru, the use of the dollar as a medium of exchange, which makes IMF policy even more inappropriate and ineffective. See Paul D. McNeils, "Stabilization, Debt and IMF Intervention: Development Trade-offs in Ecuador and Peru," *Trocaire-Irish Development Review*, **3**, 47–58, 1987.

11. *Ibid.*, 34.

12. *Ibid.*, 35.

13. IMF orthodoxy is different from the "new orthodox" macroeconomies followed by Argentina, Chile, and Uruguay in the late 1970s. See Paul D. McNelis, "The Preferential Option for the Poor and the Evolution of Latin American Macroeconomic Orthodoxies," in Thomas M. Gannon, S. J., editor, *The Catholic Challenge to the American Economy*, New York: MacMillan Publishing Company, 1987, for a description of the "Southern Cone" new orthodoxy.

14. Dornbusch, R., "Short Term Macroeconomic Policies for Stabilization and Growth." Unpublished manuscript, Department of Economics, Massachusetts Institute of Technology, 1989.

15. Edwards, S., "The International Monetary Fund and the Developing Countries: A Critical Evaluation." *National Bureau of Economic Research Working Paper No. 2909*, 1989.

16. *Ibid.*, see Tables 4 and 5.

17. It is also true that any abrupt increase in inflows, such as that from a commodity or natural resource export boom, or from the illegal exports from coca production, as well as an abrupt capital inflow, will lead to real exchange rate appreciation, and will lead to a decline in export competitiveness and production in other exports, such as manufacturing or agriculture. This experience is known as "Dutch disease."

18. Horton, S., "Labor Markets in an Era of Adjustment: Bolivia," unpublished Manuscript, Department of Economics, University of Toronto, 1989, for a description of labor market reforms in Boliva.

19. Dornbusch, R., *op. cit.*, Table 2.

20. Morales, J.A., "La Transicion de la Estabilidad al Crecimiento en Bolivia," Catholic University of Bolivia, Working Paper, Department of Economics, 1989.

21. *Ibid.*, see Cuadro 6 and the discussion on 22–23.

22. Dornbusch, R., *op. cit.*, 2.

23. McCloskey, D.N., "The Rhetoric of Economics," *Journal of Economic Literature*, **21**, 481–517, 1983.

24. —— Two Replies and a Dialogue on the Rhetoric of Economics," *Economics and Philosophy*, **4**, 150–166, 1988.

25. McCloskey, D.N., *op. cit.*, 482, 1985.

26. *Ibid*, 482–483.

27. *Ibid*, 511.
28. McCloskey, D.N., *op. cit.*, 151, 1988.
29. McCloskey, D.N., *op. cit.*, 515, 1985.
30. Rappaport, S., "Economic Methodology: Rhetoric or Epistemology," *Economic Philosophy*, **111**, 121, 1988.
31. Rosenberg, A., "Economics Is Too Important To Be Left to the Rhetoricians," *Economics and Philosophy*, **4**, 135, 1988.
32. Rappaport, S., *op. cit.*, 128.
33. McCloskey, D.N., *op. cit.*, 156, 1988.
34. Rosenberg, A., *op. cit.*, 149.
35. McCloskey, D.N., *op. cit.*, 166, 1988.
36. McCloskey, D.N., *op. cit.*, 509–510, 1985.
37. *Ibid.*, 510.
38. Maki, U., "How to Combine Rhetoric and Realism in the Methodology of Economics," *Economics and Philosophy*, **4**, 89–109, 1988.
39. *Ibid.*, 107—108.
40. Sen, A.K., *Poverty and Famines: An Essay on Entitlement and Deprivation*, Oxford: Oxford University Press, 1982.
41. Maki, U., *op. cit.*, 107.
42. Finch, D., "Let the IMF Be the IMF," *International Economy,* Jan./Feb. 1988; and Manuel Pastor, "The Effects of IMF Programs in Latin America: Debate and Evidence from Latin America," *World Development*, **15**, 249—262, 1987.
43. Rappaport, S., *op. cit.*, for a discussion of these devices.
44. Berg, A. and Sachs, J., "The Debt Crisis: Structural Explanations of Country Performance," *National Bureau of Economic Research Working Paper No. 2607*, 1988.
45. Bourguinon, F., Branson, W.H. and deMelo, J., "Adjustment and Income Distribution: A Counterfactual Analysis," Natural Bureau of Economic Research Working Paper No. 2943.
46. *Ibid*, 32.
47. McNelis, P.D. and Nickelsburg, G., "The Macrodynamic Effects of Alternative Resolution Strategies for Debtor Countries," *Revista de Analysis Economico*, **4**, 71–84, 1989.
48. Morande, F. and Schmidt-Hebbel, K., "Sovereign Debt Conversion in a Dynamic Portfolio Framework," *Revista de Analisis Economico*, **4**, 1989, pp. 51–70 and Elhanan Helpman, "The Simple Analytics of Debt/Equity Swaps," *American Economic Review*, **79**, 440–451, 1989.
49. McNelis, P.D. and Nickelsburg, G., *op. cit.*
50. In this model, I ignore two sources of inequality among households, changes in interest rate and changes in the exchange rates, which generate capital gains or capital losses for those who hold domestic or foreign financial assets. These effects are considered by Bourguinon, Branson, and de Melo,

op. cit.

51. This assumption of symmetric indexation may seem strong. Wages can be indexed symmetrically to the price level, or the rate of change of wages can be indexed symmetrically to the rate of inflation. The analysis can easily cover this latter case, when falling inflation rates lead to falling wage changes for unions currently negotiating their contracts. In order to keep the simulation simple, I analyzed first the case of price-level changes, with inflation as a part of the adjustment process rather than a steady-state phenomenon.

52. Sachs, J., *op. cit.*

53. McNelis, P.D., 1988, *op. cit.*, The Israeli stabilization plan in 1985 may have been successful, relative to the Latin American plans, precisely because the redistributive element has been part of the social pact among the government, industrialists, and unionists at the start of the stabilization plan.

Chapter 3

Part 2. Ethical and Modeling Considerations in Correcting the Results of the 1990 Decennial Census

Stephen E. Fienberg *

INTRODUCTION

The United States Decennial Census

At 10 a.m. on May 11, 1992, a federal courtroom in Manhattan was crowded with lawyers and expert witnesses, as Judge Joseph M. McLaughlin entered. "Good morning, ladies and gentlemen," he began. "This day has been a long [time] coming. I would like to say that I have been looking forward to it, but candor forbids that". Thus began the trial on the lawsuit brought by the City of New York and others seeking to overturn a 1991 Department of Commerce decision not to adjust the results of the 1990 decennial census to account for the differential undercount of Blacks and other minorities. This chapter is about the statistical models proposed for adjusting the 1990 census and the decade-long debate that surrounded their application. The debate has had both political and statistical dimensions and culminated in the presentations of 15 expert statistical witnesses at the New York City trial. The trial ended on May 27, 1992, and the judge issued his decision Against Adjustment" in April 1993.

* This chapter is based in part on material drawn from a series of papers on the adjustment of the 1990 census which appeared in *Chance* over a four year period. A major revision of this chapter was prepared while the author was at the faculty of York University, Toronto, Canada, and was supported in part by a grant from the Natural Sciences and Engineering Research Council of Canada.

April 1, 1990, marked the official date of the 1990 United States decennial census of housing and population and the 200th anniversary of the first American census. The official United States population count for 1990, announced on December 31, 1990, by Census Bureau Director Barbara Bryant, was 248.7 million people. The accuracy of this count, who these people were, where they resided, and who was missed remain questions of considerable interest, controversy, and importance. This is because social scientists and policymakers rely upon the census as a source for basic data about the nation. But even more important, the allocation of political power and federal funds depends directly on census data, and thus much rides on who is counted.

Some people think that we do a census of the United States population by literally going out and counting heads, with each person in the country being seen by an enumerator who records the relevant information on a census form. In fact, a large proportion of those included in the 1970, 1980, and 1990 censuses were never seen by an enumerator (their questionnaires were mailed out and mailed back). It is important to understand that any census, no matter how carefully it is carried out or what methodology it uses, cannot be complete, i.e., without error (see Appendix A to this chapter for information on sources of census error). Modern census organizations have introduced statistical techniques to compensate for some of the inevitable errors that occur, and some of these involve the use of sampling (see Appendix B to this chapter for examples). As a consequence, many statisticians believe that it would be more accurate to describe the census as producing estimates of the population rather than counts.[1] This is not to suggest that the statisticians at the United States Bureau of the Census are doing a poor job in producing population counts every ten years, but rather that the job they are doing is far more complex than most people understand.

The primary purpose of the United States decennial census, going back to 1790, was to provide information for use in apportioning the seats in the United States House of Representatives among the states. This is still the key statutory use to which census data are put, but information from the census has also become crucial for

* reallocation of state legislatures,
* allocation of federal funds to the states,
* the calculation of vital statistics,
* public planning and decision making,
* business planning and decision making,
* the basic sampling frame for most national sample surveys, and
* scholarly analysis.

As the content and uses of the decennial census have grown over the years, so have the costs. The 1980 census cost close to US $1.1 billion, and the 1990 census cost approximately US $2.5 billion. Due to the substantial costs associated with the census and the political outcomes that are based on data that it generates, the accuracy of the census has come under careful scrutiny by statisticians, public officials, and politicians from both major parties as well as various political action groups. Among the major concerns regarding the census over the past decade have been the inclusion or exclusion of illegal aliens from the census and the differential undercount between Blacks (and other ethnic minority groups such as Hispanics) and Whites. As we explain later in this chapter, the two issues are linked. Our focus here, however, will be only on the issue of differential undercount in the 1990 United States decennial census and the role that statistical models have played in its resolution. We explore ethical (as well as political and social) aspects of the issue, the modeling process, and the models themselves.

Alternative Views on Statistical Models in a Census Context

Stigler[2] describes probability or statistical models as falling into two classes: *micromodels* that specify the behavior of individuals or simple causes, and *macromodels* that specify the behavior of aggregates. Stigler then uses these types of models to describe currently fashionable views on statistical inference. He notes that one current school of statistical thought argues that, without the support of a validated micromodel, preferably one based on explicit random sampling, there is no basis for objective statistical inference. Here, we refer to this school as the statistical sampler. In the context of the debate over correcting the 1990 decennial census, David Freedman is the prototypic member of this school.[3] A second school treats inferences about the parameters in empirically constructed macromodels, cautiously interpreted, as fully objective as those derived from micromodels, especially given the realities of actual sampling practice in the context of large-scale data collection. A third school, not singled out for discussion by Stigler, consists of the Bayesians or subjectivists, for whom the entire process of model construction and assessment is subjective. For the present purpose, we combine members of the Bayesian school with macromodelers under the banner of *statistical modelers*, despite their differences on the objective/subjective dimension. Members of this second school active in the census debate include Kirk Wolter,[4] formerly on the staff of the Bureau of the Census, and the present author.

There are, of course, many links between the worlds of macro- and micromodels. Such linkages in the context of social phenomena can be traced back at least to the pioneering work on macromodeling of Adolphe Quetelet, whose major work took place between 1835 and 1850.[5] This is worthy of special note here, since Quetelet also played a major role in developing the modern notion of census-taking, and he viewed models as a natural component of work on censuses.

In part, the debate over correcting the 1980 and 1990 censuses for differential undercount is an academic dispute between the statistical samplers and the statistical modelers, with the samplers appearing to oppose any innovation involving statistical modeling unless it can be backed up by a micromodel with no questionable assumptions. While the position of the samplers has a seductive appeal, especially for those who believe in the objectivity and validity of mathematical models in the physical sciences, the realities of real-world sampling and data-collection are such that official statisticians and policymakers must make decisions and inferences in the absence of perfect or even plausible micromodels.[6]

A second dimension to the debate over the use of statistical models in the context of the census emanates from the fact that federal statistical agencies in the United States have long prided themselves on their independence and their ability to produce data in a neutral fashion. At a meeting several years ago, one of the speakers commented that "government statistical data should be valueless". Actually he meant value-free, and when pressed he was unwilling to admit that the wording of questions in a survey or the definition of a variable inevitably reflects a perspective that is almost certain to have a political or societal component.

Anderson,[7] for example, argues that "the earlier generations of Americans who took the census . . . framed the census inquiries and organized the data that they published in ways that mirrored their notions of contemporary social problems",[8] i.e., the census reflects the values of the society it serves and thus reinforces them. This remains true today for the United States decennial census. In Fienberg,[9] we argued that statisticians who collect data mandated by others can, at best, attempt to remain "disinterested" and impartial, but they certainly cannot uncouple their efforts from societal values and perspectives. Moreover, once produced, statistical data "enter the political fray on behalf of social interests",[10] and thus the statistician's job does not end with the production of impartial reports or data summaries. This is clearly the case when it comes to the issue of adjusting the census for differential undercount.

Despite the fact that national data cannot be value-free, we believe that, in the past, statistical agencies such as the Bureau of the Census have, by and large, done an excellent job in insulating their data from unnecessary political aspects, including the political views of those working for the agencies. Yet, even when the Census Bureau succeeds in its political "neutrality", it still collects data in the context of a mandate that comes from the political arena, i.e., from Congress or from politically-appointed federal administrators. What disturbs many statisticians about the decisions surrounding the 1990 census is the extent to which politics rather than statistics may have guided technical decisions about the census process and what would constitute the official census data. The original decision to halt the Bureau plans for correction in 1987 was made by a political appointee. The 1991 decision not to correct the census for the differential undercount (reversing a recommendation from the Bureau) was made by former Secretary of Commerce Robert A. Mosbacher, who subsequently resigned his position to lead the campaign to re-elect President George Bush. Further, at trial, counsel for the City of New York presented evidence of telephone calls from the White House to those involved in the adjustment decision. However one views the statistical issues of adjustment, there is no question that the decision not to adjust has had and will continue to have strong political overtones and implications.

Outline of Chapter

In Section 2 of this chapter, we briefly summarize what we know about the differential undercount problem and information on its magnitude in the censuses from 1940 to 1990. In Section 3, we describe in simplified terms the key statistical model proposed for use to correct the differential undercount problem and key assumptions that justify its use. Following a discussion of the political and legal controversy over adjustment for undercount in the 1980 census and in the planning for the 1990 census, we turn, in Section 4, to the recently concluded New York City trial in which the statistical models used for adjustment took center stage. In Section 5, we contrast the circumstances surrounding the use of statistical models for adjustment in the 1980 and 1990 censuses, and we present an explicit discussion of ethical issues and dilemmas raised in the public debate over the use of statistical models in the census context.

THE UNDERCOUNT PROBLEM

Some History

Concerns about the accuracy of the census counts in the United States have

Ethics in Modeling

existed almost as long as the census itself. For example, in 1791 following the first census, Thomas Jefferson wrote:

> "Nearly the whole of the states have now returned their census. I send you the result, which as far as founded on actual returns is written in black ink, and the numbers not actually returned, yet pretty well known, are written in red ink.
>
> Making a very small allowance for omissions, we are upwards of four millions; and we know that omissions have been very great".[11]

Almost 100 years later, General Francis A. Walker,[12] Superintendent of the United States Censuses of 1870 and 1880, writing in the *Journal of the American Statistical Association* about the undercount of Blacks in the 1870 census, elicited one of the earliest statistical proposals for adjustment for the undercount from H. A. Newton[13] and H. S. Pritchett,[14] both of whom used the method of least squares to fit a third-degree polynomial to census data for 1790 to 1880 and then measured the undercount for 1870 as a residual from the fitted curve.[15]

The Magnitude of the Undercount Problem Since 1940

Beginning with the 1940 census, the Bureau of the Census estimated the size of the undercount by race, using a technique known as demographic analysis (described in greater detail below). From 1940 through to 1980, these estimates showed improved overall accuracy of the census, but the difference in the rate of undercount for Blacks and non-Blacks has remained roughly constant, somewhere between 5% and 6%. The actual figures are given in Table 1. We note that it is the *differential* undercount between Blacks and non-Blacks that is important when we come to assess census accuracy, since census figures are typically used to divide resources (e.g., political representation and dollars) among groups in the population.

In fact, there are various versions of demographic analysis estimates. In the top panel of Table 1, we give the "traditional" version based on the methods in use through the 1980s. During 1990, the Census Bureau introduced a major revision to these estimates associated with the rule for identifying the race of children of mixed marriages. This change predictably produced a decrease in the differential undercount. These revised figures are given in the bottom panel of Table 1.

For the 1990 census, the preliminary demographic estimate of the overall undercount was 1.9%, about 50% higher than the demographic estimate of undercount in 1980. Furthermore, the differential undercount for Blacks was

6.4% or 5.3% depending on the methodology used. This appears to be the largest differential undercount since the Census Bureau began to measure it back in 1940.

Table 1. Estimated Net Census Undercount from 1940 to 1990 as Measured by Demographic Analysis

Year	Black	White	Difference	Overall undercount
1940	10.3%	.1%	5.2%	5.6%
1950	9.6%	3.8%	5.8%	4.4%
1960	8.3%	2.7%	5.6%	3.3%
1970	8.0%	2.2%	5.8%	2.9%
1980*	5.9%	0.7%	5.2%	1.4%
1990**	7.4%	1.0%	6.4%	1.9%
1990 Revised Estimates*				
1940	8.4%	5.0%	3.4%	5.4%
1950	7.5%	3.8%	3.8%	4.1%
1960	6.6%	2.7%	3.9%	3.1%
1970	6.5%	2.2%	4.3%	2.7%
1980*	4.5%	0.8%	3.7%	1.2%
1990**	5.7%	1.3%	4.4%	1.8%

*The figures for 1980 are based on the assumption that 3 million undocumented aliens were living in the United States at the time of the census.

**The figures for 1990 are based on the assumption that 3.3 million undocumented aliens were living in the United States at the time of the census.

***Revised to reflect 1990 change in methodology in the classification of mixed race births.

Sources: R. E. Fay, J. S. Passel, J. G. Robinson, and C. D. Cowan. *The Coverage of Population in the 1980 Census*. Bureau of the Census United States Department of Commerce. Washington D.C., 1988.; Bureau of Census press release, June 13, 1991, and unpublished Bureau of Census reports.

It is important to keep in mind that the 1.9% figure does not mean that the census correctly counted over 98% of the United States population (as suggested by various public officials and repeated by opponents of correction such as Freedman and Navidi[16]). Rather, it represents the *net* undercount, which can be thought of as the difference between the actual undercount (consisting of missed individuals or omissions) and the overcount (erroneous enumerations and duplications). The 1980 undercount of 1.3% translated into an overall or gross error rate of 7%, and the figure for 1990 of in excess of 10%.

Table 2. Estimated Census Undercount for Racial/Ethnic Groups as Measured by the 1990 Post-Enumeration Survey

Group	Undercount	Differential undercount
Total	2.1%	
Non-Black	1.7%	
Black	4.8%	3.1%
Hispanic	5.2%	3.4%
Asian	3.1%	1.0%
American Indian	5.0%	2.9%

Source: Bureau of the Census press release, June 13, 1991.

Until 1980, only limited attention was directed towards the count of the Hispanic population and how it is affected by underenumeration. In fact, Hispanics are lumped together with Whites in the comparisons presented in Table 1. Moreover, the vast majority of the illegal alien population resident today in the United States is Hispanic, and this makes the interpretation of the 1980 figure problematic. While it has been widely believed that the undercount for Hispanics and Blacks, especially for those in similar socioeconomic conditions, is similar, i.e., the "cause" of the differential undercount problem is basically socioeconomic and cultural, not racial,[17] it was only in the 1990 census that direct national estimates for the differential undercount for Hispanics and for various racial groups were first calculated. Table 2 contains a summary of these estimates. Note that the estimated differential undercount for Blacks reported here is less than that reported in

Table 1 and is based on different methodology.

STATISTICAL TECHNIQUES FOR ESTIMATING THE UNDERCOUNT

Two Basic Methods

Two quantitative techniques have been used to estimate the undercount at a national level: demographic analysis and the dual-system or capture–recapture technique.

Demographic analysis combines birth, death, immigration, and emigration records with other administrative records to carry forward the population from one census to the next, deriving an estimate of the overall population size and thus the undercount. The methodology can be used to provide population and undercount figures by age, race, and sex, but only at a national level. Demographic analysis cannot be used to provide reliable state, regional, and local estimates, principally because of the absence of accurate data on migration.

The "completeness" of the data used in the demographic analysis represents the most crucial underlying model assumption. The oldest birth cohorts are not well represented in official birth records, and there are questions regarding the accuracy of both birth and death records over time with respect to the identification of race. This is of special concern for the Hispanic population, which was not separately identified in the official report of the demographic analysis of the 1980 and 1990 censuses. Furthermore, the official figures on immigration do not reflect illegal immigrants. We note that the 3 million illegal residents added into the 1980 demographic analysis was less than the number of illegal aliens who applied for and were granted legal status following the 1986 Immigration Act Revision (and thus could provide documentation that they had entered the United States prior to 1982). In 1990, the Bureau estimated that there were 3.3 million illegal residents in the United States. These estimates of illegal residents are quite "soft" and rest on suspect survey data and a number of assumptions, but there is general agreement that a high proportion of the estimated number of illegal residents in both censuses was Hispanic. Finally, we note that the last component needed for demographic analysis, emigration data, is virtually non-existent.

We can view demographic analysis as a statistical method, and it is therefore natural for us to ask about estimates of the variability or uncertainty associated with the demographic analysis estimate of the differential undercount. Unfortunately, attempts to estimate the uncertainty in the demographic estimates are quite crude and not well documented. From a statistical modeling perspective, the demographic analysis approach is based on a macromodel that has natural links to basic microdata. This is not how many demographers view the method, however, and it has only been in connection with the demographic analysis for the 1990 census that they began to prepare "judgmental" confidence bounds for the estimates.[18]

The other basic method for undercount estimation has two different names — the dual-system method or the capture-recapture method.[19] In this approach, those included in the census are matched with those found in a second source (e.g., a random sample of the population or a list based on administrative records), and this information is used to produce an estimate of those missing in both sources and thus an estimate of the undercount in the original census. Fienberg[20] provides an annotated bibliography for the capture–recapture literature, with special emphasis on its use in the context of census undercount correction. We give a more detailed description in the following subsection.

The Dual-System Model: Heuristics

Dual-system estimation uses two sources of information to arrive at a better estimate of the population than could be obtained from one source alone. The procedure is an old one (dating back to the nineteenth century), widely accepted among statisticians, and has been used for a host of different practical population estimation problems. It is most familiar in the context of estimating the size of wildlife populations, where it is called the "capture–recapture" technique. For example, a marine biologist wishing to estimate the number of fish in a lake can do so by twice attempting to catch and count each fish in the lake. In order to keep track of which fish were counted the first time, the biologist uses the simple expedient of marking each fish caught before releasing it. As the fish are caught the second time, the biologist can then determine, for each fish, whether it had been counted in the first measure of the population. Information from these two counts is then used to derive a more accurate estimate of the size of the fish population than either of the two counts alone can provide.

Suppose the biologist counted 200 fish the first time and 150 fish the second time, and that of the 150 fish counted the second time, 125 bore marks indicating that they had been among the 200 fish counted the first time. There are thus three categories of fish that have been counted: fish caught both the first time and the second time (125 in this example), fish caught the first time but not the second (75 in this example, or the total number of fish caught the first time minus the number of marked fish recaught the second time), and fish caught the second time but not caught the first time (25 here, equaling the total number caught the second time minus the number of marked fish caught the second time). The total number of fish in the three classes is 225. Note that all 225 have been directly observed by the marine biologist, and that this number exceeds the number of fish observed in either of the two counts.

The task is to proceed from this information to an estimate of the total number of fish in the population, including an estimate of a fourth category of fish, those not caught by the biologist either time. This can be done as long as the two fish counts are independent random samples. A random sample by its nature permits us to estimate the incidence of any observable characteristic in the larger population from which the sample is drawn. In this case, the characteristic we are interested in is that of having been captured in the first count. Examination of our random sample — the second count — showed that 125 out of 150 fish, or 5/6 of the sample, had been captured in the first count. Generalizing from the sample, we can conclude that 5/6 of the total fish population in the lake was captured in the first count.

Having learned that 5/6 of the total population was covered by the first fish count, we are in a position to estimate the true fish population on the basis of the available information. The total number of fish counted the first time, 200, is 5/6 of the total population, i.e.,

$$200 = (5/6) \hat{N}$$

where $\hat{N}$ is the estimate of true population size, N. To arrive at our estimate, a little high school algebra suffices:

$$\hat{N} = (6/5) (200)$$

$$= 240.$$

Of this estimated 240 total population, 225 have been observed in one or the other or both of the two counts. Thus, we can infer that there are 15 fish in the population that were not counted either time. This is our estimate of the number of fish in the fourth category of the population.

A demographer wishing to measure the size of a human population can follow the same procedure to arrive at a more accurate estimate than any single count could provide. By counting a population twice, and by comparing the two counts to see how many people are in each of the three observed population classes (those counted the first time but not the second, those counted both times, and those counted the second time but not the first), an accurate estimate of the true population size can be derived, including people not counted either time.

For example, suppose we are trying to estimate the number of people living on a block in the Bedford-Stuyvesant neighborhood of Brooklyn. The first count of the population could be a list of the names and addresses of people counted in the census. Assume that 400 people were counted by the census on this particular block. For the second count of the population, let us imagine a specially commissioned, independent count. In the hypothetical Bedford-Stuyvesant block, the second count will surely include some of the same 400 people counted by the census. We determine how many people were counted both times by comparing the list of names and addresses in the first count with the list of names and addresses in the second count, and matching those that are the same. Suppose that, through matching these two lists, we find that 250 people out of 300 counted in the second count on this block were also included in the census. The three observed categories of the population, and their sizes, are thus the 250 counted both times, the 150 counted the first time but not the second, and the 50 counted the second time but not the first. If the second count is a random sample of the population of the block as a whole, then 250 out of 300, or 5/6 of the whole population of the block were counted in the census. The fraction of members of the second count included in the census is an estimate of the fraction of the total population of the block counted in the census. Finally, we can estimate that the total population is the number of people counted in the census, or 400, divided by 5/6. This yields an estimate of the total population of 480 people. Thus, the estimated census undercount in this hypothetical block is 80 people out of 480, or 1/6 of the population.

Since the undercount rate varies among different kinds of blocks, separate undercount rates must be estimated for each of a sample of blocks. We would not, for example, expect our dual-system estimate of the undercount rate of a block in Bedford-Stuyvesant to tell us very much about the undercount in an expensive suburb. Accordingly, in selecting the sample, one starts with a list of all the blocks or equivalent-size rural units in the United States, and groups the blocks into categories, or "strata", according to their demographic characteristics. Such characteristics might include racial composition, proportion of home-owners to renters, and average household size. Rates of undercount are determined for each stratum by measuring the undercount of a sample of blocks within the stratum by dual-system estimation. The undercount rate of these blocks is then applied to the other blocks throughout the country in the stratum. The actual correction of the census counts consists of adjusting the raw census count for each stratum to compensate for the estimated undercount in that stratum.

The Dual-System Model: More Formal Details

Suppose there are x_1+ individuals counted in the census and the second source contains x_{+1} individuals, x_{11} of whom match with individuals in the census. Then the traditional capture-recapture estimate for the overall population size, N, is

$$\hat{N} = x_{1+} x_{+1} / x_{11}. \tag{1}$$

A closely related way to think of the estimation process is in terms of a 2 x 2 table of counts (this is the reason for the subscripted notation above). The rows of the table correspond to being included or not in the census and the columns correspond to being included in the second source. Then we can display the data as follows:

		Sample		
		In	Out	Total
	In	x_{11}	x_{12}	x_{1+}
Census	Out	x_{21}	x_{22}	x_{2+}
	Total	x_{+1}	x_{2+}	N

$$(2)$$

We observe x_{1+} from the census and x_{+1} from the second source. By matching the two sources, we find x_{11}, and thus get x_{12} and x_{21} by subtraction. Finally, x_{22} (and thus x_{2+}, x_{+2}, and N) is unknown and we estimate it. If being counted in the census is independent of being counted in the second source, then we expect

$$x_{11} / x_{12} = x_{21} / x_{22} \tag{3}$$

and thus we use as our estimate of the unobserved x_{22}:

$$\hat{x}_{22} = x_{12} x_{21} / x_{11} \tag{4}$$

Finally, we add this estimate for the missing cell to the three observed counts to get our estimate of the population size, N:

$$\hat{N} = x_{11} + x_{12} + x_{21} + \hat{x}_{22}$$
$$= x_{1+} + x_{21} + x_{12} x_{21} / x_{11} \tag{5}$$

which is identical to the estimate in equation (1). The final version of the estimate of N in equation (5) has three components:

1. the basic count, x_{1+},
2. the count of those seen in the second source but not in the census, x_{21},
3. the projection from the model of the projected number of those missed in both, $x_{12}x_{21}/x_{11}$.

Provided that the matching of records from the second source with those from the census is done without error, the second of these components represents actual individuals who were not included in the census but should have been.

This same basic methodological approach has long been linked to the analysis of census data, e.g., see Tracy[21] and Shapiro,[22] and it has been used to adjust the count in the Australian census.[23] The United States Bureau of the Census introduced the use of a sample matched to the census records for coverage evaluation in connection with the 1950 decennial census. This dual-system approach has evolved and is currently used in the United States to measure the census undercount nationwide by taking an independent count of the populations of each of a large number of blocks nationwide and then matching the results of these counts to the results of the census for those blocks. This second count is known as a "post-enumeration survey" or PES. The census and the PES are the two counts used to estimate the population of each of the blocks.

The modification of the dual-system estimate to deal with the use of a sample from the census results rather than the census itself, including the use of a complex sample survey as the second source is described by Wolter[24] and Cowan and Malec.[25] In essence, we replace equation (1) by

$$\hat{N}^* = x_{1+}^* x_{+1}^* / x_{11}^* \tag{6}$$

where $x_{1+}^* = x_{1+} - II - EE$, II is the weighted number of people in the census with insufficient information to be matched, EE is the estimate of the number of erroneous enumerations in the census, x+1* is the weighted number of people for the sample used as a second source, and x_{11}* is the weighted number of people in both the census and the sample. The main difference between equations (1) and (6), aside from the adjustment for II and EE, is that the revised 2 x 2 table entries corresponding to equation (6) are sample counts weighted up to represent population totals.

Implementing Dual-System Estimation in a Census Context

The census modification of this technique can be used directly at the national level as well as at state and substate levels. This modified dual-system estimation approach was used in conjunction with both the 1980 and 1990 censuses as part of the post-enumeration survey (PES) program. We describe the different approaches used in these two censuses.

In the 1980 PES, the Census Bureau took a sample of 110,000 households from the census, selected in clusters of approximately 10 housing units per enumeration district, and matched it to the households in the April and August Current Population Survey (CPS), each containing approximately 84,000 households. The Bureau produced estimates of the undercount for the United States as a whole, as well as for all 50 states and several large local areas.[26] Substantial controversy surrounded the subnational undercount estimates that emanated from the 1980 PES.[27] In 1990, the Census Bureau revised its PES methodology in order to better estimate the differential undercount. Instead of relying upon a survey designed for a totally different purpose (the CPS), the Bureau designed a new PES directly linked to census geography. The Bureau implemented this approach in a 1986 test of adjustment-related operations in Los Angeles,[28] and, following the completion of the original enumeration in July 1990, it gathered survey information from the inhabitants of approximately 5,000 blocks across the nation, matching the information collected with the results of the original enumeration for these blocks. In all, the PES involved checking on the occupants of 165,000 households nationwide. The 1990 PES-based estimates were the focus of the recent New York City lawsuit described below. In Appendix C of this chapter, we include a detailed description of the PES-based procedure used to implement the capture–recapture methodology in 1990 and to produce corrections for the differential undercount for every census block in the nation.[29]

Dual-System Model Assumptions and Their Empirical Base

As with the demographic analysis method, the dual-system estimation method is based on a set of assumptions linked to a very explicit statistical model. The three assumptions most widely discussed are:

(i)   **perfect matching.** The individuals in the first list can be matched with those in the second list, without error.

(ii)   **independence of lists.** The probability of an individual being included in the census does not depend on whether the individual was included in the second list.

(iii)   **homogeneity.** The probabilities of inclusion do not vary from individual to individual.

Perhaps the greatest problem with the dual-systems approach as it was used in conjunction with the 1980 census was the rate of matching errors (the failure of assumption (i) above). There are two kinds of matching errors: false matches and false non-matches. If two different individuals are erroneously matched, then x_{11} is erroneously increased and x_{12} and x_{21} decreased. The net effect of false matches is to decrease the second and third components in the estimate of the population size from equation (5) and thus to underestimate the population size. If individuals in the sample cannot be matched with their records in the census, then x_{11} is erroneously decreased, and x_{12} and x_{21} are increased, leading to overestimation of the population size.

The matching problem was discussed at length in the analyses reported by Fay *et al.*[30] and was the focus for a major research program following the 1980 census and the development of new computer-based matching algorithms. Jaro[31] reports on the new methods and their accuracy. Approximately 75% of the 1990 PES records were matched using these algorithms, and special features of the redesigned PES reduced other difficulties associated with the matching process. Schenker[32] describes the imputation methodology used to allocate the residual unmatchable records in 1990.

The failure of assumption (ii) is known as *correlation bias*. In the presence of positive correlation bias (being missed in the census is positively correlated with being missed in the second list), the traditional estimator of equations (1) and (5) tends to underestimate the actual population size but yields an improvement over the unadjusted value.[33] While it is widely believed that there was positive correlation bias in the 1980 PES estimates,[34] there is little solid empirical evidence on the issue. The only direct statistical method available to measure the extent of correlation bias involves a generalization of the dual-system approach to multiple lists and the estimation techniques developed for multiple-capture problems.[35] This alternative is complex and has only been used in an indirect test census context[36] and not as part of the full-scale census process. Thus, we have a major assumption associated in our model that most believe to be false, and we have insufficient empirical evidence to replace it by a more reasonable one. As part of the 1990 census correction process, the Bureau produced several indirect estimates of correlation bias using techniques such as those suggested in Wolter,[37] all of which reaffirmed what had widely been assumed: that correlation bias appears to be positive. Thus, the PES-based procedure only partially corrected for the differential undercount.

Many questioned the homogeneity of the probabilities of inclusion in the 1980 census and they often argued that violation of assumption (iii) also contributed to correlation bias. While evidence on this assumption is at best anecdotal, the stratification structure that was built into the design for the 1990 PES reduced concern at the Census Bureau about heterogeneity. As part of the followup to the PES, the Bureau examined several indirect measures of the impact of residual heterogeneity. Darroch *et al.*,[38] have proposed an alternative approach to the estimation heterogeneity based on multiple lists that may be of value in the future.

Smoothing Dual-System Estimates

A special feature of the procedure proposed by Ericksen and Kadane[39] in connection with adjustment of census counts for 1980 was the "smoothing" of the PES-based dual-system estimates for each of 66 areas using a multiple regression model. Much of the critique of Freedman and Navidi[40] focused on this model, the choice of predictor variables for inclusion in it, and the appropriateness of the standard regression assumptions. This debate over the regression models continued in Ericksen, Kadane, and Tukey,[41] and in Freedman and Navidi.[42]

In planning for the 1990 PES, the Census Bureau did extensive analysis of possible predictor variables that would be available as part of the census enumeration data for use in a new "smoothing" multiple regression model. This research included simulation studies using 1980 data,[43] and it led to a partially prespecified set of multiple regression models, one for each of several regions of the country, with one group of variables being included with certainty and others selected from a second group based on their predictive power.

A second feature of the regression model involved the "presmoothing" of the estimated variances of the dual-system estimates, using a simple regression model, prior to the use of these variances in the regression model itself. This feature was introduced after extensive analysis of PES results from the 1988 census dress rehearsal, and it is the subject of considerable debate.

Finally, the smoothed dual-system estimates were combined with the original dual-system estimates using an empirical Bayes procedure, which generated final estimates that were a weighted average of the two. In essence, each "raw" dual-system estimate (with presmoothed variances) was shrunk towards its predicted value from the multiple regression model, with the degree of shrinkage being weighted by factors depending on their variances (cf., Step 7 of Appendix C to this chapter.)

ADJUSTMENT OF THE UNDERCOUNT IN THE 1990 CENSUS: THE POLITICAL OVERLAY

The 1980 Dispute and the Bureau's Research Program

Prior to the 1980 census, there was extensive discussion in the statistical community regarding the advisability of adjusting the census counts to correct for the undercount.[44] A decision was made shortly before the official reporting deadline, in December 1980, not to adjust the results for the anticipated differential undercount,[45] although Kadane[46] and others contend that this decision was in fact made prior to the taking of the census. A lawsuit was filed on census day by the city of Detroit requesting that the 1980 census be adjusted for the undercount, and this action was followed by 52 others, 36 of which requested adjustment. One of these cases, brought by the state and city of New York gained considerable attention, with a large number of statisticians testifying for and against adjustment.[47]

The New York lawsuit, known as *Cuomo* v. *Baldrige*, ultimately went to trial in January 1984 (an earlier trial ruling was overturned by the United States Supreme Court), but the judge did not issue his opinion until December, at which time he ruled that no adjustment need be made. He argued that, because statisticians and demographers can and do disagree on the reliability of an adjustment of the 1980 census, it would be inappropriate for the court to substitute its judgment for that of the experts at the Census Bureau. The articles by Ericksen and Kadane[48] and by Freedman and Navidi[49] reflect some of the statistical arguments presented in court in this case, and Conk[50] gives an excellent historical perspective of the social–political dimensions of the 1980 adjustment debate.

Simultaneously with these activities, the Census Bureau launched a major research program to improve the methodology used for census adjustment, and it commissioned the Committee on National Statistics (CNSTAT) of the National Research Council to establish a Panel on Decennial Census Methodology, whose charge included the review of the census undercount research program. The panel's 1985 report[51] outlined the basic issues that needed to be addressed in the adjustment research program. Subsequently, the panel reviewed the proposed methodology developed by the Census Bureau staff for adjustment in 1990 and its implementation in two separate pretests. This methodology was based on a newly designed post-enumeration survey and the use of dual-systems estimation and was designed to overcome problems with the PES/dual-systems approach used in the 1980 census as noted above. In two separate later reports, in 1986 and 1987, the CNSTAT panel made a positive assessment of the technical feasibility of the planned adjustment methodology. In addition, at two separate 1987 congressional hearings, Bureau officials and independent statistical experts testified on issues relating to the technical feasibility of adjustment.[52]

A Halt to Adjustment Planning

Then, in October 1987, Under Secretary Robert Ortner of the Department of Commerce, in which the Bureau of the Census is located, announced that the 1990 Census would not be adjusted for the differential undercount. This decision was widely criticized by many in the statistical community.[53] Unlike the situation in 1980, when there was a serious split in opinion among statisticians knowledgeable about census-taking and adjustment methodology, the vast majority of statistical experts, inside and outside of the Census Bureau, concluded that the methodology proposed for use in 1990 was statistically sound, and they urged that the Bureau be allowed to proceed with its implementation. A special congressional hearing was held in March 1988, focusing on the technical feasibility of the adjustment methodology and reviewing the decision not to plan for adjustment.[54] The hearing provided documentation on the deliberations at the Bureau prior to the announcement and substantiates the charges of political interference, including the following:

- there was a consensus among the statisticians in the Undercount Research Staff and others in the Statistical Standards and Methodology Division that the adjustment methodology had been successfully implemented in the Los Angeles Census Pretest and "that adjustment was technically sound and feasible".

- in May 1987, the Bureau's Undercount Steering Committee recommended proceeding with plans for adjustment, and the Bureau Director concurred in this recommendation.

- the Bureau's plans were overruled by political officials in the Commerce Department.

The decision not to plan for adjustment would clearly benefit the Republicans, whose administration made the decision in opposition to professional statistical advice. Census Director John Keane, a political appointee of the Republican administration, resisted pressure to take a politically expedient position on adjustment, and it was Under Secretary Ortner who attempted to put a scientific gloss over what had become an intensely political decision. To make matters worse, individuals at the Census Bureau attempted to "rewrite history" by belatedly setting to paper a rationale that they claimed to have been true back in the spring of 1987 for not having proceeded with plans for adjustment.

Resuming Planning for the PES

In October 1988, the Commerce Department decision was challenged in a new lawsuit brought against the federal government and its officials by the City of New York and other state, county, and local governments. They alleged that the decision not to adjust was arbitrary and not based on technical grounds. The trial was scheduled to take place in July 1989, but the parties reached a last-minute settlement and entered into a Stipulation and Order supervised by the court. The October 1987 decision was withdrawn, and the Commerce Department agreed to consider the issue of whether or not to carry out a statistical adjustment of the 1990 census *de novo*. The Commerce Department also agreed to announce standards for adjustment publicly prior to the census, and it stipulated that all census data released or published prior to an adjustment decision would carry a disclaimer, saying that the results were subject to possible correction. Finally, the Secretary of Commerce agreed to appoint an independent panel of eight experts to advise on adjustment-related activities, including four members chosen from a list of experts proposed by the plaintiffs in the lawsuit.[55]

The panel was appointed in September, held a series of meetings, and prepared a report on a preliminary version of the proposed adjustment guidelines, which set out standards by which the quality of the PES would be assessed as a means for adjusting the 1990 decennial census population counts. In October 1989, a Department of Commerce official announced that the Department was relying on its recently appointed panel to provide advice on adjustment and that it would not provide support to any other panel to review these issues. This controversial announcement had the effect of cutting off the Committee on National Statistics' Panel on Decennial Census Methodology, which had been studying the issue of adjustment since early 1984. The panel had been expected to resume activities where it had left off just prior to the filing of the New York City lawsuit in 1988. This Department of Commerce decision led to the disbanding of one of the most visible and independent sources of advice for census-adjustment and coverage evaluation advice and did little to suggest the openness of the adjustment process to input from the professional statistical community.

A final version of the adjustment guidelines was published in the *Federal Register* on March 15, 1990. Typical of the comments submitted by statisticians was that of Morris H. Hansen of Westat, Inc., and a former Associate Director of the Bureau of the Census:

I am especially concerned over the fact that the proposed guidelines appear to reflect a predetermination by the Department of Commerce that an adjustment will be disadvantageous and will not be made. . . [The guidelines document ignores] the fact that the methodology of the decennial census has been constantly improved over the past two hundred years, through experience and research. Presumably it can be further improved, perhaps by an adjustment . . . [and it] reminds me of the statements of a former patent commissioner . . . that all useful inventions had been made and no additional worthwhile inventions can be expected.

The plaintiffs in the 1988 lawsuit claimed that the guidelines had no technical content, provided no rules for decision on the question of correction, and were biased against correction. They filed a motion in federal district court in which they requested a declaratory judgment and a court order declaring the guidelines promulgated by the Department null and void, holding that statistical adjustment of the census to correct for the differential undercount does not violate the Constitution, and ordering defendants to adjust the census, unless they set forth by June 1, 1990, a set of guidelines establishing standards and procedures for evaluating accuracy and demonstrating that the original enumeration is more accurate. Counsel for the Department of Commerce argued that all of the claims of the plaintiffs were without merit, that the claims for a declaratory judgment and for a supplemental Order were "not ripe for review", and that it was not proper for the Court to entertain "such imagined grievances at this juncture".

A month later, Judge Joseph McLaughlin issued his ruling on the motion supporting some of the plaintiffs' requests and denying others. On the matter of the guidelines, he noted that they were the "bare minimum" required under the original Stipulation and Order and thus denied the plaintiffs' request to have them set aside. He also observed that "because defendants have chosen to contribute adequate but minimal performance to satisfy their obligations at this stage, defendants clearly incur a heavier burden to explain why no adjustment was made in the event the Secretary [of Commerce] elects to proceed with an actual enumeration [without adjustment]". Further, he warned the defendants that "intentional inaction will not be tolerated. Defendants are expected, and indeed required, to honor their solemn commitments embodied in the Stipulation. . . . [Defendants] are on notice, if it is not already clear, that backdoor attempts to evade their commitment will not be countenanced".

On the matter of the constitutionality of adjustment, the judge held that the United States Constitution "is not a bar to statistical adjustment" and that "[t]he concept of statistical adjustment is wholly valid, and may very well be long overdue". He went on to observe: "that said, it does not follow that any and all forms of statistical adjustments will be sanctioned. . . . Whether it has been done legally and constitutionally can only be determined after the Secretary has decided how he wishes to adjust, if at all".

The 1990 Census Enumeration

In the midst of the legal dispute over adjustment guidelines, the detailed process of census-taking began. On approximately March 23, 1990, households all over the United States received 1990 census forms in the mail, and these forms were to be completed and returned to the Bureau by mail. On April 27, when the door-to-door phase of the enumeration began, only 63% of the forms mailed out in March had been completed and returned. This was far below the expected rate of return of 70% that the Census Bureau had planned for in its budget, and the Bureau was forced to request US $110 million of additional funding from Congress in order to attempt to solve the problems caused by this shortfall. Extra enumerators were hired, and the deadline for the door-to-door canvass was extended from June 7 to June 30. The Post-Enumeration Survey (PES) began on June 25 and continued throughout the summer. Final followup and checking of results of the original enumeration proceeded throughout this period. On December 31, 1990, Bureau Director Barbara Bryant announced that the official United States population count for 1990 was 248.7 million, and she released state counts to serve as the basis for the reapportionment of the United States House of Representatives.

The Decision Not to Adjust

On April 18, 1991, the Census Bureau released preliminary results from the PES, confirming the earlier reports from the demographic analysis of a significant differential undercount affecting Blacks. The PES data and detailed information on adjusted counts were subject to subsequent scrutiny by Bureau statisticians, by members of the Special Secretarial Advisory Panel, and by a number of consultants. Their reports were due at the Bureau in mid-June, giving Bureau staff and the Department of Commerce exactly one month to reach a final decision.

The Special Secretarial Advisory Panel was split 4–4 on their adjustment recommendations, with the plaintiffs' experts recommending adjustment and the government's experts opposing it. On June 21, the internal Census Bureau Undercount Steering Committee, which reviewed the final evidence from PES evaluation studies, voted 7–2 in favor of adjustment. Bureau Director Barbara Bryant (who had been appointed the previous year) supported the majority recommendation to adjust and forwarded it with her blessing to the Commerce Department. Later she was quoted as saying that she believed an adjustment was warranted because it would improve the counts for a majority of states and for the places where a majority of the nation's population lives.

Even before the Commerce Department announced its decision, allegations of political interference were in the air. On the morning of July 15, 1991, the *New York Times* carried a story of calls made from the White House staff to members of the Special Secretarial Advisory Panel. At 3:30 p.m. that afternoon, Secretary of Commerce Robert A. Mosbacher announced that "[a]fter a thorough review, I find the evidence in support of adjustment to be inconclusive and unconvincing. Therefore, I have decided that the 1990 census counts should not be changed by a statistical adjustment". The Commerce Department decision not to adjust, announced by Secretary Robert A. Mosbacher on July 15, was accompanied by 5 thick volumes of supporting material and that analyses weighed in at an estimated 15 pounds. These detailed materials attempted to martial a strong case against adjustment and emphasized that the capture–recapture methods appropriate for fish and animal populations require special justification for use with human populations (thus ignoring the extensive literature on the topic extending back over 40 years that is referred to earlier).

Preparing for Trial

The primary focus of attention quickly shifted from Washington back to New

York City and the courtroom of Judge Joseph M. McLaughlin. At the request of New York City and the other plaintiffs, the judge reopened the case in light of the Secretary's decision not to adjust, set an early November trial date for the lawsuit brought originally in 1988, and ordered an expedited process of evidential discovery. The government appealed this order, arguing that depositions were unnecessary since the Secretary had already provided the full basis for his decision in his 5-volume report. This had the effect of delaying all official trial-related activities until October 27, when Supreme Court Judge Anthony Kennedy denied the appeal.

In November, as the taking of depositions was just getting under way, plaintiffs sought to depose Secretary Mosbacher. The government immediately moved to block this attempt on the familiar legal grounds that allowing the deposition of high government officials in the course of their duty is action that should be allowed only in the most compelling situations. Judge McLaughlin ultimately ruled for the government, but the delay led to yet another postponement of the trial. The taking of depositions continued throughout the winter and, after a series of additional delays, Judge McLaughlin set a trial date in May 1992.

The New York City Trial

The 1990 New York City census adjustment trial began on May 11, 1992. The trial lasted three weeks, and it consisted primarily of testimony from a large number of expert statistical witnesses: for the plaintiffs, there were Eugene Ericksen, Kirk Wolter, John Rolph, Leo Estrada, Barbara Bailar, Franklin Fisher, Bruce Cain, John Tukey, and Stephen Fienberg; for the defense, Peter Bounpane, Robert Fay, Paul Meier, Leo Brieman, Kenneth Wachter, and David Freedman. Five of these witnesses also served as members of the Special Secretarial Advisory Panel (Ericksen, Wolter, Estrada, and Tukey from the plaintiffs and Wachter from the defendants). In his opening statement, Judge McLaughlin essentially laid out the ground rules for the case:

The plaintiffs requested that the court hold a trial. The defendants objected to conducting a trial, arguing that because the case arose under the Administrative Procedure Act (APA) the scope of the court's review should be limited to the administrative record. On February 18th of this year the court . . . concluded that the evidentiary hearing is necessary . . . for several reasons.

First, there have been several allegations, some of them serious, that the administrative record is a self-serving . . . compilation of documents, assembled for the purpose of strengthening the defendant's litigation position. The court therefore must consider the integrity of the administrative record to ensure that these documents, and only these documents, were considered by the Secretary.

The second purpose of the hearing is to enhance the court's understanding of the record, replete as it is with references to technical jargon and arcane procedures. To this end, the testimony of expert witnesses hopefully will prove helpful. Their testimony should also assist the court in determining whether the Secretary considered all the relevant factors in making his decision.

Finally, the court is fully aware of the important constitutional questions this case presents. In their opening statements, attorneys for the two sides of the litigation identified their positions on the issues and the roles of the experts they planned to call. Counsel for the plaintiffs recalled the 30-page report submitted to Secretary Mosbacher by Census Bureau Director Barbara Bryant in which she had recommended the adoption of the adjusted data on the basis of her detailed and careful review of the Bureau's evaluation of the PES results. He then summarized the extensive research program carried out by the Census Bureau throughout the decade of the 1980s leading up to the PES. All of this information, he claimed, was seemingly ignored by the Secretary in his "official decision" and little or no information on it was included in the "so-called" administrative record prepared after the fact, under the supervision of trial counsel from the Department of Justice.

Our evidence will show that [the decision] is utterly groundless, that all its convoluted obscurity is a sham, dressed up to look like scientific analysis but utterly fallacious, root and branch. Indeed, your Honor, if it were a brief submitted by counsel in this court, counsel would be subject to serious rebuke for serious and patent misuse of the record. We will focus both on what the decision omits and what it includes. What it omits is critical, and what it includes is erroneous. . . . Your Honor, the arbitrariness and capriciousness with which the secretary acted is enough to overturn his decision and reestablish the decision of the expert agency [the Bureau of the Census]. Although I think we do not need to demonstrate any motive on the part of the secretary, we intend to show your Honor that there is ample evidence of the real rationale at work in this decision, although not apparent in the decision document.

In response, counsel for the defendants noted that, because this case is one under the APA, the Secretary's decision cannot be disturbed unless it is shown to be arbitrary and capricious and the only remedy, in the event that the Court finds the Secretary's decision was unreasonable, is a remand to the Secretary with instructions to reconsider his decision. He then referred to the guidelines which, he noted, provided that the census was to be considered more accurate than an adjustment unless it could be shown that adjusting the 1990 decennial census would result in a more accurate picture of the proportional distribution of the population of the United States on a state-by-state basis. Then he came to the methodology of adjustment:

I'm going to summarize what the evidence will show about how adjustment is done. And I think that's important, especially in this case, your Honor, because in some ways, adjustment is like a salami. You might like the way salami tastes, but you might not eat it if you saw how it was made.

He then proceeded to give an overview of the PES and the adjustment methodology, emphasizing that the Bureau relied throughout the PES on statistical models that depended upon critical and unjustifiable assumptions.

Any attempt to summarize in a few paragraphs 2,600 pages of trial transcript covering three weeks of highly detailed testimony will of necessity result in oversimplification. In Fienberg,[56] we describe the major topics covered in the questioning of the expert witnesses. Here we simply focus on a number of broad themes that arose repeatedly and on which the experts for the plaintiffs (P) and the defendants (D) can be contrasted. These included:

1. P's experts disagreed not merely with the Secretary's decision but specifically with the details presented and the evidence it relied upon. D's experts, while agreeing with the decision, rarely referred to the detailed contents of his official report, although when they did they agreed with his assessment.

2. P's experts had all worked on the technical aspects of the adjustment methodology covered in their testimony and had all published directly or indirectly on that work. D's experts were somewhat mixed in this regard and were critics of the methodology, at least for adjustment purposes.

3. P's experts said that while departures from assumptions were important, they were relative and had to be understood in context;

on balance the PES-adjusted counts were still superior to the raw enumeration. D's experts said that departures produce biases and as a result the PES and an adjustment based on it were fatally flawed.

4. P's experts relied almost completely on the work done by the Bureau in conducting and analyzing the PES up through the Secretary's decision. D's experts chose to disregard or criticize much of the Bureau's work and relied heavily on their own analyses, most of which were completed many months after the Secretary's decision.

5. All but one of P's experts did all of their work on a *pro bono* basis. D's outside experts were paid substantial fees by the government for their work.

Final briefs were filed at the end of July and the judge issued his decision in the case 9 months later on April 13, 1993. He ruled that the Commerce Department's decision not to adjust "not be disturbed", but he granted the plaintiffs' request that the adjusted data be released for others to use.

REFLECTIONS UPON STATISTICAL MODELS AND STATISTICIANS IN THE CENSUS CONTEXT

Perspectives on the Use of Statistical Models for Adjustment in 1980 and 1990

The suggestion by some that the issues and evidence regarding adjustment of the census in 1990 are much as they were in 1980 is far from the mark. For the 1980 census, the decision not to adjust was made in advance of the reporting of the official December 1980 report date by the Director of the Bureau, Vincent Barabba, who had the authority for the decision delegated to him by the Secretary of Commerce. Thus the plaintiffs in *Cuomo* v. *Baldrige* were challenging the decision of the expert agency and its judgment. For the 1990 census, the decision not to adjust was made by the Secretary, counter to the recommendation of the Census Bureau and its Director, Barbara Bryant. Plaintiffs in the 1990 New York City lawsuit were thus explicitly saying that the Court and the Secretary should defer to the Bureau's judgment; the government was in effect challenging the judgment of its own agency.

During the 1990 New York City trial, the defendants contended that the statistical models used for adjustment in 1990 were similar to those that were proposed for use in 1980 and that "reasonable people could and did disagree as to their appropriateness". While it was true that the general nature of the statistical models used in 1990 for dual systems estimation and for smoothing were similar to those suggested in the 1980 litigation, the technical and empirical support for their use was far different.

Prior to the 1980 census, dual-system estimation was viewed mainly as part of the Census Bureau's coverage program, and little effort was expended by the Bureau on plans for a possible adjustment. In fact, after Director Barabba's decision not to adjust, the Bureau made no effort to use the PES results to infer adjustments to the enumeration counts at the block level, something that would be required if adjusted data were to be used for reapportionment. There had been only a limited amount of research on dual-system estimation during the 1970s, and this work received public scrutiny only during the 1980s after its publication.[57] A panel of the Committee on National Statistics (CNSTAT) reviewed plans for the 1980 census, but its report (Panel on Decennial Census Plans, 1978) had only a limited discussion of methodology for adjustment and it made no recommendation regarding its use in 1980.

In part as a result of the lawsuits over adjustment in 1980, a very different set of circumstances unfolded during the planning for 1990. The Bureau itself identified adjustment methodology as a major focus of that planning effort, and this activity led to a total redesign of the PES, substantial field tests, and planning for full-scale implementation in the census itself. Much new research was initiated by the Bureau in connection with plans for adjustment, and dozens of papers were published in professional journals on the topic, many of which we have cited here. Many papers explore the departure from assumptions that have been viewed as fundamental to the dual-system and smoothing methodologies linked to the PES. Interest in the issue of adjustment was especially strong within the American Statistical Association, and there were several sessions on the topic at each of its annual meetings beginning in the mid-1980s. In fact, a Congressional Hearing on the topic was held by the House Subcommittee on Population and Census as part of the 1988 ASA Annual Meetings, and a number of statisticians appeared as witnesses. The journal *Survey Methodology* devoted the better part of its 1988 issues to the topic of census methodologies, and most of the papers dealt with aspects of adjustment. The *Journal of the American Statistical Association* published two special discussion papers on the topic of adjustment[58] which includes a special section on adjustment of the 1990 census in a 1993 issue. Finally, we recall the recommendations of a new CNSTAT panel in support of the Bureau's plans for adjustment.

In many ways the evaluation effort surrounding the development and implementation of 1990 PES may have made it the most studied survey in the history of the Census Bureau. Much of the work surrounding this effort has appeared in the statistical profession's most prestigious refereed journals. While all of this activity has certainly provided a much better foundation for the use of statistical models to adjust the results of the 1990 decennial census than was the case in 1980, there is still no unanimity within the profession on the advisability of their use. Nor would one expect such unanimity, since it rarely occurs in connection with the use of any other statistical methodology.

Ethical Issues

One might argue that the Bureau of the Census did a heroic job in getting the 1990 population counts accurate at a national level to within a few percent and that we should simply applaud their achievement. The problem with such an argument is that the few percent error is the net error (at a national level where lots of errors balance out) and it disguises how the overcounts and undercounts distribute themselves, resulting in much higher levels of net undercount for Blacks and Hispanics, groups who have looked to the federal government for special protection to address their problems. Thus the major ethical issue we need to consider as it relates to the differential undercount is equity. We are not directly invoking the principle of equality here, i.e., that everyone should count equally (unlike in the original census of 1790 when slaves were counted as fractions), but rather the notion of equity that presumes that all groups should be protected from harm. In the present instance, harm is measured in lost seats in Congress and in state legislatures, in lost dollars in federal allocations, and so on. If we know that Blacks and Hispanics are undercounted relative to Whites, resulting in harmful consequences, and we know that, on average, adjustments can be made that will move us closer to the truth, then should not the statistician feel compelled, ethically, to use statistical models to correct this inequity?

Despite the fact that the evidence substantiating the differential undercount comes from samples and the use of statistical models, there is a professional consensus that the differential undercount exists and adversely impacts upon Blacks and Hispanics. Thus, we do begin with the knowledge that if we do nothing at all, the census counts will be incorrect and in a form predictable in advance, especially in the aggregate. Thus, we argue that the current methods are intrinsically unfair. What is in dispute is the magnitude of the problem and the uncertainty as to the impact of new coverage improvement programs.

There is a related technical argument—one rooted in some of the assumptions linked to the dual systems model for population estimation—that using the method to correct for undercount will move the original counts towards the unknown true counts. Taking into account the total error structure of the PES-based dual systems estimates (including the empirically demonstrable departures from assumptions), I personally believe that the adjusted counts are closer to the true counts than are the counts from the original enumeration. That I must use professional judgment and knowledge to reach this conclusion does not undercut its validity.

The second major ethical issue relates to the political arena in which the census drama is being played out. Were adjustment merely a technical exercise of little practical import, then we would expect to see the technical arguments for and against (and the related empirical support) published in refereed scholarly journals until some form of closure on the issue was reached. As we described in the preceding section, much of this drama has been played out in an intensely partisan setting, and the very fact that there has been a high-quality and public technical debate at all makes the adjustment issue of special interest. Many of the statistical arguments and methods used in the 1990 census adjustment debate have now been published, e.g., see the articles in the special section on census undercount in the September 1993 issue of the *Journal of the American Statistical Association*. Nonetheless, because census adjustment is an inherently political issue the executive branch, the federal courts, and Congress have not allowed the matter to be settled in professional journals. The ethical dilemma we statisticians face comes from the need to distance ourselves from the political positions that have been staked out by other players, while at the same time gaining entree to the public debate. In addition, some statisticians have to address a potential conflict of interest between the technical involvement and their political values and goals.

For the "independent" Special Secretarial Advisory Panel, the very process of panel selection was partisan in nature, with half of the members coming from a list provided by each side in the lawsuit. Members of each group discussed the implication of technical choices with those who had something to gain (politically or otherwise) from the outcome of the panel's deliberations. Several panel members had access to confidential information regarding census data, processes, and outcomes that may have had value to them in other professionally-related activities, especially subsequent to the decision on adjustment. Finally, there were allegations that the Department of Commerce attempted to manipulate the resources that it had made available to the Panel (an allegation confirmed by a ruling from Judge McLaughlin), and to restrict access to confidential information on the part of some panel members. Such a set of circumstances is ripe for concern about ethics and conflict of interest.

This second issue is especially relevant for those statisticians who became involved as statistical experts, either as members of the Special Secretarial Advisory Panel or as expert witnesses in the New York City adjustment lawsuit. What becomes clear from an examination of the role of statisticians and other professionals as expert witnesses in legal proceedings more broadly[59] is that the very nature of the United States adversarial legal system draws the expert witness away from neutrality and objectivity. The process begins with the briefing of the expert by counsel who invariably presents the facts of the case from the perspective of the client. Moreover, access to various types of data and information is often a function of the party with whom the expert is working. As an expert's involvement with a case grows, so too may friendship with counsel. Fisher[60] reminds us that "Particularly because lawyers play by rules that go beyond those of academic fair play, it becomes insidiously easy to see only the apparent unfairness of the other side while overlooking that of one's own side. Continuing to regard oneself as objective, one can slip little by little from true objectivity."

Addressing the broad range of litigation areas involving statistical testimony, Meier[61] describes these extra-technical dimensions of the statistician as an expert witness quite well:

[T]he professional integrity of the expert witness and, through him, of the profession that he represents is not well protected by the courts and hardly at all by counsel. But before we assume too readily that simple morality and personal ethics will be an adequate substitute, we should reflect for a bit on . . . corrupting influences. . .

First, there is the fact that the expert witness is playing someone else's game and, inevitably, has to accept the rules as he finds them. His instructor in these matters is, of course, his client's counsel, and the witness is ill-equipped to resist the role of adversary when his lawyer thrusts it upon him.

Among the most difficult of the corrupting influences to deal with is what I call *aggrandizement*. . . . Long ignored and treated with contempt in literature and in the courts, the statistician has been elevated to Olympian levels. . . . He will be tempted to ignore or minimize those qualifications that he might emphasize in an academic setting, he may fail to emphasize schools of thought other than his own, and he may lay claim to overly broad scope for the inferences he draws.

Meier goes on to describe a host of other additional influences added by the adversarial system including bribery, flattery, co-option, and personal views. Meier also advocates the use of personal and professional codes as a way to defend the integrity of statistical testimony.

Among the specific issues of relevance to those statisticians testifying in the New York City census adjustment trial are the following:

- the objectivity of Census Bureau employees, or those dependent on the Bureau financially, as expert witnesses when their jobs relate to the success of the Bureau's parent Department as a party to the litigation.

- the objectivity of others serving as expert witnesses (fee for service or public service?) — e.g., the government witnesses were paid substantial fees in connection with their litigation-related work, as was one witness for the plaintiffs.

- playing the lawyers' game by allowing oneself to make judgments with certainty of the quality of estimates and potential outcomes in 1990 from fragmentary evidence.

- the absence as witnesses of statisticians at the Bureau who favored the use of the adjusted counts.

- the absence as witnesses of statisticians receiving research grants (Joint Statistical Agreements) from the Bureau — was there an implicit threat to the continuation of their funding if they had participated as witnesses against the government?

The first two of these relate to issues of conflict of interest, either explicitly or implicitly.

This discussion of ethical issues is far from inclusive, but it should offer insights into the extent that statisticians and other professional modelers confront major ethical dilemmas when they work on important problems of public policy, such as whether or not to adjust the decennial census to correct for differential undercount.

Epilogue

The story of correcting the census in 1990 really began with the court cases associated with the dispute over the differential undercount in the 1980 census. More than a decade of research and controversy has passed, and the statistical community as well as the Federal District Court have considered the use of statistical models in a process design to adjust the census counts for the differential undercount. It is now more than two years since the results of the 1990 census were officially released, but the story is barely complete. In December, 1992, Director Barbara Bryant announced that the Census Bureau would not adjust the intercensal estimates for the differential undercount as estimated by the PES, a decision made in the waning days of the Republican administration and viewed by many as statistically unwise and politically driven. Then on April 13, 1993, Judge McLaughlin issued his opinion in the New York City lawsuit:

> Having thus passed the guidelines, the Court concludes that the Secretary's conclusions under each guideline cannot be characterized as arbitrary and capricious. The breadth of the guidelines left the Secretary enormous discretion. Plaintiffs have made a powerful case that discretion would have been more wisely employed in favor of adjustment. Indeed, were this Court called upon to decide the issue *de novo*, I would probably have ordered the adjustment.* However, it is not within my providence to make such determinations. The question is whether the Secretary's decision not to adjust is so beyond the pale of reason as to be arbitrary and capricious. That far I cannot go.

> Balance against the slight residuary interest that the defendants may have in the confidentiality of the block level data is the public's interest in full access to judicial proceedings, especially where as here, the dispute has sparked so much public interest. Because I believe that the balance weighs heavily in favor of disclosure under these circumstances, I vacate the protective order and permit plaintiffs to use and to release to the public the computer tapes containing the adjusted block-level counts.

An extensive planning activity is already well underway at the Census Bureau, looking to make changes in the census for the year 2000. Members of both the House of Representatives and the Senate continue to express an ongoing interest in these activities. In October 1991, both the House and the

Senate passed and President Bush signed a bill, known as the "Decennial Census Improvement Act of 1991", authorizing a new three-year independent study of alternatives for the decennial census of the year 2000, to be conducted by the Committee on National Statistics (CNSTAT) at the National Academy of Sciences. Representatives Tom Sawyer and Harold Rogers addressed the June 1992 opening meeting of the panel, and they indicated that, while part of the panel's efforts should focus on ways to produce a more cost-effective and accurate census for 2000, the panel should also look beyond 2000 so that more substantial changes can be properly tested and evaluated. A related CNSTAT panel, on the technical aspects of census methodology. In their preliminary 1993 reports, both CNSTAT panels appear to have accepted what Judge McLaughlin characterized as the inevitability of adjustment in 2,000. Thus we can expect a continuing political interest in and dimension to census planning for 2000 and beyond.

NOTES AND REFERENCES

1. Citro, C.F. and Cohen, M.L., eds., *The Bicentennial Census. New Directions for Methodology in 1990*. Report of a Panel of the Committee on National Statistics. Washington D.C.: National Academy Press, 10–11, 1985.
2. Stigler, S.M., "The role of probability models in statistical inference in 19th century Europe". *Bulletin of the International Statistical Institute, 47th Session, Invited Papers*, Book 1, 157–162, 1989.
3. Freedman, D.A., "Policy forum: Adjusting the 1990 census". *Science*, **252**, 1233–1236, 1991.
4. Wolter, K.M., "Policy Forum: Accounting for America's uncounted and miscounted". *Science*, **253**, 12–15, 1991.
5. Stigler, S.M., *The History of Statistics: The Measurement of Uncertainty Before 1900*. Cambridge, MA: Harvard University Press, Cambridge, MA: 1986.
6. See Freedman, D.A. and Navidi, W.C., "Should we have adjusted the census of 1980?" (with discussion). *Survey Methodology*, **18**, 3–74, 1992, and Kruskal, W.H. Introduction, in *Measurement Error in Surveys* (P. Biemer, R. Groves, L. Lyberg, N. Mathiowetz, and S. Sudman, eds.). Wiley, New York, xxiii–xxxiii, 1991.
7. Anderson, M.J., *The American Census: A Social History*. New Haven, CT: Yale University Press, 1988.
8. *Ibid*, xi.
9. Fienberg, S.E., "An adjusted census in 1990?" *Chance*, **2:3**, 23–25, 1989.
10. Prewitt, K., "Public statistics and democratic politics". In Wm. Alonso and P. Starr, eds., *The Politics of Numbers*. New York: Russell Sage Foundation, 261–274, 1987.
11. Jefferson, T., Letter to David Humphreys, in *The Papers of Thomas Jefferson*,

Charles T. Cullen, ed., Princeton University Press, **22**, 62, 1986.

12. Walker, F.A., "Statistics of the colored race in the United States". Publications [later *Journal*] *of the American Statistical Association*, **2**, 91–106, 1890.

13. Newton, H.A., "Note on President Walker's article on statistics of the colored race". *Publications* [later *Journal*] *of the American Statistical Association*, **2**, 221–223, 1891.

14. Pritchett, H.S., "A formula for predicting the population of the United States". *Publications* [later *Journal*] *of the American Statistical Association*, **2**, 278–286, 1891. [Reprinted from *Transactions of the Academy of Science, St. Louis*, 1891.]

15. Stigler, S.M., "The Centenary of JASA". *Journal of the American Statistical Association*, **83**, 583–587, 1988.

16. Freedman, D.A. and Navidi, W.C., *op. cit.*, 74, 1992.

17. Hogan, H. and Wolter, K., "Measuring accuracy in a post-enumeration survey". *Survey Methodology*, **14**, 99–116, 1988.

18. Passel, J. S., Das Gupta, P. and Robinson, J.G., "Evaluation of demographic analysis: development of confidence intervals". Paper presented to Census Advisory Committees, October 1987.

19. Bishop, Y.M.M., Fienberg, S.E. and Holland, P.W., *Discrete Multivariate Analysis: Theory and Practice*. Cambridge, MA: M.I.T. Press, 1975, Chapter 6.

20. Fienberg, S.E., "The trial". *Chance*, **5:3–4**, 28–38, 1992.

21. Tracy, W.R., "Fertility of the population of Canada". Reprinted from *Seventh Census of Canada*, 1931, Vol. 2, Census Monograph no. 3. Ottawa: Cloutier, 1941.

22. Shapiro, S., "Estimating birth registration completeness". *Journal of the American Statistical Association*, **45**, 261–264, 1949, and "Recent testing of birth registration completeness in the United States". *Population Studies*, **8**, 3–21, 1954.

23. Choi, C.Y., Steel, D.G. and Skinner, T.J., "Adjusting the 1986 Australian census count for underenumeration". *Survey Methodology*, **14**, 173–189, 1988.

24. Wolter, K.M., "Some coverage error models for census data". *Journal of the American Statistical Association*, **81**, 338–346, 1986.

25. Cowan, C.D. and Malec, D., "Capture-recapture models when both sources have clustered observations". *Journal of the American Statistical Association*, **81**, 347–353, 1986.

26. Fay *et al.*, *op. cit*, 1988.

27. See the extended debate between Ericksen, E.P. and Kadane, J.B., "Estimating the population in a census year: 1980 and beyond (with discussion)". *Journal of the American Statistical Association*, **80**, 98–131, 1985 and Freedman, D. and Navidi, W.C., "Regression models and adjusting the 1980 census (with discussion)". *Statistical Science*, **1**, 3–39, 1986 and its continuation in Ericksen, E.P., Kadane, J.B. and Tukey, J.W., "Adjusting the 1980 census of population and housing". *Journal of the American Statistical Association*, **84**, 927–944, 1989 and Freedman, D.A. and Navidi, W.C., "Should we have adjusted the

census of 1980?" (with discussion). *Survey Methodology*, **18**, 3–74, 1992.

28. See Diffendal, G. "The 1986 Test of Adjustment Related Operations in Central Los Angeles County". *Survey Methodology*, **14**, 71–86, 1988; Hogan, H. and Wolter, K., "Measuring accuracy in a post-enumeration survey". *Survey Methodology*, **14**, 99–116, 1988; and Mulvey, M.H. and Spencer, B.D., "Total error in the dual system estimator: the 1986 census of Central Los Angeles County". *Survey Methodology*, **14**, 241–263, 1988.

29. See Hogan, H., "The 1990 post-enumeration survey: An overview". *American Statistician*, **46**, 261–269, 1992, and Wolter, K.M., *op. cit*, 1991.

30. Fay *et al.*, *op cit*, 1988.

31. Jaro, M.A., "Advances in record-linkage methodology as applied to matching the 1985 census of Tampa, Florida". *Journal of the American Statistical Association*, **84**, 414–420, 1989.

32. Schenker, N., "Handling missing data in coverage estimation with application to the 1986 Test of Adjustment Related Operations". *Survey Methodology*, **14**, 87–98, 1988.

33. Hogan, H. and Wolter, K., *op. cit*, 1988.

34. See Freedman, D. and Navidi, W.C., *op. cit.*, 1986, and Hogan, H. and Wolter, K., *op. cit*, 1988.

35. See Bishop, Y.M.M., Fienberg, S.E. and Holland, P.W., 1975, *op. cit.*, and Fienberg, S.E. "The multiple-recapture census for closed populations and the 2^k incomplete contingency table". *Biometrika*, **59**, 591–603, 1972.

36. Zalavsky, A.M. and Wolfgang, G.S., "Triple system modeling of census, post-enumeration survey, and administrative list data". *Proceedings of the Section on Survey Research, American Statistical Association*, 668–673, 1990.

37. Wolter, K.M., "Capture-recapture estimation in the presence of a known sex ratio". *Biometrics*, **46**, 157–162, 1990.

38. Darroch, J.N., Fienberg, S.E., Glonek, G.F.V. and Junker, B.W., "A three-sample multiple-recapture approach to census population estimation with heterogeneous catchability". *Journal of the American Statistical Association*, 88, 1993, in press.

39. Ericksen, E.P. and Kadane, J.B., *op. cit*, 1985.

40. Freedman, D. and Navidi, W.C., *op. cit*, 1986.

41. Ericksen, E.P., Kadane, J.B. and Tukey, J.W., *op. cit*, 1989.

42. Freedman, D.A. and Navidi, W.C., *op. cit*, 1992.

43. Isaki, C.T., Schultz, L.K., Diffendal, G.J. and Huang, E.T., "On estimating census undercount in small areas". *Journal of Official Statistics*, **4**, 95–112, 1988.

44. Keyfitz, N., "Information and allocation. Two uses of the 1980 census". *American Statistician*, **33**, 45–50, 1979.

45. Mittroff, I.I., Mason, R.O. and Barabba, V.P., *The 1980 Census: Policymaking and Turbulence*. Lexington, MA: D.C. Heath, 1983.

46. Kadane, J.B., "Book review of Mittroff, Mason, and Barabba". *Journal of the*

American Statistical Association, **79**, 467–469, 1984.

47. See the editorial prologue to Ericksen, E.P. and Kadane, J.B., *op. cit,* 1985.
48. Ericksen, E.P. and Kadane, J.B., *op. cit,* 1985.
49. Freedman, D. and Navidi, W.C., *op. cit,* 1986.
50. Conk, M.A., "The 1980 census in historical perspective". In Wm. Alonso and P. Starr, eds., *The Politics of Numbers.* New York: Russell Sage Foundation, pp. 155–186, 1987.
51. Citro, C.F. and Cohen, M.L., *op. cit,* 1985.
52. Wallman, K.K., "A tale of two cities: Act III", *Chance,* **1:2**, 48–52, 1988.
53. For example, see Bailar, B.A., "Statistical practice and research: The essential interactions". *Journal of the American Statistical Association,* **83**, 1–8, 1988.
54. Wallman, K.K., "A tale of two cities: Act III". *Chance,* **1:3**, 55–57, 1986.
55. Fienberg, S.E., *op. cit,* 1989.
56. Fienberg, S.E., *op. cit,* 1992.
57. For example, see Cowan, C.D. and Malec, D., "Capture-recapture models when both sources have clustered observations". *Journal of the American Statistical Association,* **81**, 347–353, 1986, and Wolter, K.M., *op. cit,* 1986.
58. See Ericksen, E.P. and Kadane, J.B., *op. cit.,* 1985 and Mulry, M.H. and Spencer, B.D., "Total error in PES estimates of population (with discussion)". *Journal of the American Statistical Association,* **86**, 839–863, 1991.
59. Fienberg, S.E., ed., *The Evolving Role of Statistical Assessments as Evidence in the Courts.* (Committee on National Statistics and the Committee on Research on Law Enforcement and the administration of Justice, National Research Council) New York: Springer, 1989(b).
60. Fisher, F.M., Statisticians, econometricians, and adversary proceedings. *Journal of the American Statistical Association,* **81**, 277–286, 1986.
61. Meier, P., "Damned liars and expert witnesses". *Journal of the American Statistical Association,* **81**, 269–276, 1986.
62. Fienberg, S.E., *op. cit.,* 1992.
63. Wolter, K.M., *op. cit.,* 1991.
64. Ericksen, E.P. and Kadane, J.B., *op. cit.,* 1985.
65. Fienberg, S.E., "Political pressure and statistical quality: an American perspective on producing relevant national data". *Journal of Official Statistics,* **5**, 207–221, 1989(c).
66. Coale, A.J., "The population of the United States in 1950 by age, sex, and color - a revision of census figures". *Journal of the American Statistical Association,* **50**, 16–54, 1985.
67. Conk, M.A., "The 1980 census in historical perspective". In Wm. Alonso and P. Starr, eds., *The Politics of Numbers.* New York: Russell Sage Foundation, 155–186, 1987.
68. Fienberg, S.E., "Undercount in the United States decennial census". In S. Kotz and N. L. Johnson, eds., *Encyclopedia of Statistical Sciences,* Supplemental Volume, New York: Wiley, 181–185, 1989(a).

69. Fienberg, S.E., "An interim report". *Chance*, **3:1**, 19–21, 1990(a).
70. Fienberg, S.E., "Back to court again". *Chance*, **3:2**, 32–35, 1990(b).
71. Fienberg, S.E., "The judge rules and the PES begins". *Chance*, **3:3**, 33–36, 1990(c).
72. Fienberg, S.E., "Commerce says no." *Chance*, **4:3**, 44–52, 1991(a).
73. Fienberg, S.E., "A full-scale judicial review approaches". *Chance*, **4:4**, 22–24, 29, 1991(b).
74. Fienberg, S.E., "Bibliography on capture-recapture modelling with application to census undercount adjustment" *Survey Methodology*, **18**, 143–154, 1992.
75. Panel on Decennial Census Plans (1978). *Counting the People in 1980: An Appraisal of Census Plans*. Committee on National Statistics, National Research Council. National Academy of Science, Washington, DC.

APPENDIX A: SOME INFORMATION ON SOURCES OF CENSUS

ERRORS

The following list provides some examples of sources of error in the census enumeration process in 1990 and indications of their possible implications for the accuracy of the census enumeration counts:

(i)   *The appropriateness of census concepts and definitions:* During the 1980s there was considerable discussion and empirical study of "housing units" and "families." These official census definitions for 1990 are often difficult to apply in practice, especially by respondents and enumerators in minority communities where nontraditional physical and family structures may be the norm. Individuals and households are often missed as a consequence. Also, individuals who are connected with multiple households may well be counted multiple times.

(ii)   *Census day vs. census operations:* Few if any people were enumerated at their household locations on April 1, 1990. (Recall that this was a Sunday!) Yet many components of the census assume that people are correctly reported in the exact location where they resided on April 1.

(iii) *Use of household respondents:* One person completes the census form for each household. This person serves as a filter for information on others in the household and household characteristics. Errors that enter into census data as a result of the "filter" applied by the household respondent are extremely difficult for anyone to correct.

(iv)   *Problems of understanding connected with the instructions and questionnaire wording:* Many individuals in the population may have had difficulty in interpreting the instructions on the census form. For example, the census form is quite explicit that college students should be counted where they go to school. Nonetheless, many respondents believe that the students should be counted at home, and as a result, college students are often double-counted.

(v) *Recording error:* Individuals completing the census questionnaire may check the wrong answer inadvertently.

(vi) *Enumerator error:* When a household fails to return the mailed-out census questionnaire, a census enumerator may complete the questionnaire as a result of a visit to the household location. Enumerators may record different information than that supplied by the household respondent.

(vii) *Coding and processing errors:* After a census questionnaire is completed and returned to the Bureau, the information must be transferred to a computer file and subsequently processed and checked for consistency. Errors creep into the official files despite the new technologically-based methods used in 1990 for data capture.

(viii) *Geo-coding errors:* These errors occur at various stages in the process from the compilation of initial mailing lists as part of the TIGER system all the way through coding. As late as September 1990, complaints from local communities of missed housing units were attributed by the Census Bureau to geo-coding problems—some of these, presumably, involve geo-coding errors in census records. Despite all of the "corrections" that occurred throughout the various forms of data review, residual geo-coding errors will place people in the wrong census blocks.

(ix) *Fabrication:* Every census there are anecdotal reports of enumerators "curbstoning," i.e., fabricating questionnaires for real and often imaginary households. The Bureau has a variety of methods to catch such fabrication, but inevitably a substantial number of fabricated questionnaires are included in the official census results.

(x) *"Last resort" information:* Some completed census questionnaires go through an extensive review process and are edited for accuracy following reinterviews. Other questionnaires are actually filled out as a last resort by enumerators without their directly seeing or interviewing any household occupants. In such cases, enumerators use information from mailmen, neighbors, or building managers that may be inaccurate, incomplete or even intentionally false. Preliminary reports suggest that procedures used in 1990 produced many more questionnaires partially completed by enumerators based on last resort information.

(xi)   *Parolees and Probationers:* The Bureau introduced a new coverage
         improvement program to count parolees and probationers using address
         information on file with state and local correction departments. A large
         proportion of the individuals added into the count by this program were not
         interviewed and might well have been living and already recorded in the
         census at addresses other than those on the lists used by the Bureau for this
         program. The final "resident" population total released in December was
         just under 3 million above the total released in August as the preliminary
         count. A substantial proportion of this increase was, according to statements
         by Bureau Director Barbara Bryant, based on the results of this program.

(xii)   *Imputation:* In 1990, as in past censuses, the Census Bureau used various
         forms of statistical estimation (based either explicitly or implicitly on
         statistical models) to "fill in" missing data. This process of filling in is
         usually referred to as imputation (see Appendix B for further details.)

APPENDIX B: EARLY CENSUS USES OF SAMPLING AND

STATISTICAL MODELS

During the course of the discussions regarding undercount in the 1980 and 1990
censuses, many arguments were raised against the use of adjustment or correction
procedures, of the sort described in this paper. The simplest of these was the "purist"
argument, which claims that Title 13 of the United States Code (the enabling
legislation for the decennial census) refers to population counts and does not permit
the use of statistical models and sampling. Others have taken issue with this legal
interpretation and note that there is a tradition of using both sampling and statistical
modeling in the context of the decennial census. For example, the census
questionnaire comes in two versions, the long and the short form. Only a sample of
the nation receives the long form, and we infer results from the sample to the nation
for those items that are on the long form alone.

In 1990, as in past censuses, the Census Bureau used various forms of statistical
estimation (based either explicitly or implicitly on statistical models) to "fill in"
missing data. This process of filling in is usually referred to as *imputation*. If the
Bureau had no information about the occupancy status of a housing unit, it imputed
a status to it, i.e. it was either occupied or vacant. If a unit was imputed as occupied,
or if the Bureau otherwise believed it to be occupied, then it imputed a number of
people to the household, as well as their characteristics. The current method of
choice for imputation, known as the "sequential hot-deck" procedure, selects a
housing unit proximate in processing as the donor of the characteristics. The
statistical "model" underlying the imputation method is that the housing units are
likely to be neighboring and have similar characteristics. In 1980, the Bureau added
3.3 million people to the census through imputation. Of these, 762,000 were added

into housing units for which the Bureau had no knowledge whether or not the units were occupied. The numbers for 1990 were much smaller as a result of "last resort" programs that introduced additional nonsampling errors (see Appendix A).

In 1970, two coverage improvement programs involving sampling were used to add 1.5 million people to the census. In the National Vacancy Check the Bureau checked a sample of 13,500 housing units previously reported as vacant to see if they actually were. Based on the sample results that 11% were in fact occupied, the Bureau proceeded to add over 1 million people to the census. In the Post-Enumeration Post Office Check, sampling was used at a secondary follow-up stage to add almost 0.5 million people to rural southern areas covered by the study.

APPENDIX C: A PRIMER ON THE CENSUS CORRECTION PROCESS

This appendix gives a step by step description of the 1990 census correction procedure using the modified dual system formulation in Section 3.3. This account is reproduced from Fienberg[62] and has been adapted from material in Wolter.[63]

The census correction process was built around the PES, which is an integral part of the following process:

Step 1. An area probability sample of about 5,000 blocks was selected. The sampling unit, the "block", was essentially a city block in urban and suburban areas and a well-defined piece of geography in rural areas.

Step 2. The 5,000 sample blocks generated two probability samples of people, the "population" or "P sample" and the "enumeration" or "E sample". The E sample consisted of all persons in the 1990 official enumeration (OE) in those blocks and was used to estimate erroneous enumerations. The P sample consisted of all persons counted in an independent enumeration of the blocks conducted some time following the OE. Together, the two samples comprised the PES.

Step 3. The P-sample persons were matched to lists of persons counted in the OE. The match was based on name, address, and various characteristics. The objective was to determine which P-sample people were counted in the OE and which were not. The initial phase of matching was performed by an automated computer matching program that matched about 75% of the P sample to their corresponding OE.

Step 4. Each E-sample enumeration was either matched or not matched to a P-sample enumeration. E-sample enumerations were ultimately designated as correct or erroneous.

Step 5. The data were then screened for any incomplete, missing, or faulty items. Values for all of the missing data were estimated by statistical imputation techniques.

Step 6. Estimates of the total population were calculated within each of 1,392 poststrata, based in part on the characteristics of the P- and E-sample people. The poststrata, based on demographic (age, racial or ethnic group, gender, owner-renter) and geographic variables, were mutually exclusive and spanned the entire United States population. The modified dual-systems estimator of total population within a poststratum can be rewritten from equation (6) of Section 3.3 to take the form:

$$\hat{N} = (X - \hat{E}) \hat{p}$$

where X denotes the actual population count achieved in the OE, $\hat{E}$ denotes an E-sample-based estimator of the total erroneous enumerations in the OE, and $\hat{p}$ denotes a P-sample-based estimator of the proportion of the total population that was enumerated in the OE.

Step 7. The 1,392 ratios or "raw" adjustment factors, N/C, where C was the count from the OE, were "smoothed" to reduce the sampling variability. "Smoothed" adjustment factors were obtained by shrinking the raw adjustment factors towards a predicted value from a multiple regression model, with the degree of shrinkage determined by the variance of predicted value and the inherent sampling variability in the raw factor. This smoothing and shrinkage is a modification of the approach proposed by Ericksen and Kadane[64] for use with the 1980 PES.

Step 8. The smoothed adjustment factors were applied to the OE, block by block, for each of the approximately 7 million blocks in the country.

Chapter 3

Part 3. The Role of Models in Managerial Decision Making — Never Say the Model Says

Vincent P. Barabba

As I have been more of a model user than a model builder, I will talk about the use of models in managerial decision making and the role of ethics in these endeavors.

I was reminded that I had actually been credited with creating a modeling law when I was doing a literature review and came across a copy of a TIMS commentary by Herb Blitzer of Eastman Kodak. In his commentary, Herb stated, "Consider Barabba's Law: 'Never say the model says'".[1] Barabba's law came into being at Eastman Kodak in reaction to a concern I had about how models were being oversold to our management. At that time, in an effort to get access to scarce resources, some of our model builders were overpromising, either by commission or omission, what the model could actually accomplish. Too often I was hearing, in response to some of the most complex and difficult of questions, "Of course, the model could answer that question. And at the push of a button!"

Well, even though being armed with the security of Barabba's Law relative to modeling was encouraging, I was more concerned about bringing ethics into the discussion. I soon found out, however, that I was not alone in my lack of experience and knowledge on the subject of modeling and ethics. I found some solace from Greenberger, Crenson, and Crissy as they noted in their 1976 book on models in the policy process, "At present, there are few professional restraints in the modeling field to curb excessive behavior. There is little in the way of an organized modeling profession to provide any authoritative standards of performance or codes of conduct for modelers".[2]

The very fact that RPI held an Ethics in Modeling Workshop in 1989 presumes: (1) that in the field of modeling there is widespread awareness that there are ethical considerations in modeling, and (2) that there is a willingness to address them directly. Now, although the workshop brought some validity to the first presumption, I am not sure of the validity of the second presumption. Nevertheless, I will suggest a few meaningful and practical ways to enhance ethical behavior. Indeed, my suggestions will focus more on causing the users of models to be more aware of what they are using and on offering a framework within which open communication among all members of the modeling team can be conducted.

This framework is crucial, for it encourages diverse views to be expressed and consensus to be reached. This is of importance, of course, because usually each of us has different views of a given topic or a different perception of reality. Again, this is not new news. Consider the allegory of the cave in Plato's *Republic*, where the acquisition of human knowledge is likened to the shadows and reflections seen on cave walls. People and objects passing outside the cave may have a different appearance because of the way light shines into the cave at any given point in time and different perspectives depending on where the observer is seated.

It would be naive for people sitting inside the closed walls of the cave to interpret the shadows as the real world. In my mind, the message of the allegory still holds. While we strive to move closer to reality with our models, we must always be conscious of the fact that the knowledge we produce is also based on the images or reflections we see on the walls of our own caves.

The framework I discuss is designed to encourage, as well as allow, the direct involvement of different views of reality. It is my belief that the diverse involvement of both builder and user will enhance the ethical behavior of all members of the modeling community because, quite simply, it is very difficult to behave unethically when everyone else is aware of and understands what you are doing — and can speak out about your actions.

Before I go any further, I feel I should outline the definition of ethics within which I have developed my thinking. Without getting too philosophical, my view is that ethics can be treated either as a moral issue or as a code of conduct. The dictionary offers these definitions:

1. A set of moral principles or values.
2. A theory or system of moral values.
3. The principles of conduct governing an individual or group.

Just to be clear, I would like to note that models themselves cannot be ethical or unethical; only the people who build or use them can have these characteristics. Models can be perceived as reasonable or unreasonable, depending on the communication between the builder and the user and on the preconceived notions of the user (or reviewer).

While I am not proposing that we establish a code of ethics for the modeling community, I'd like to note that others like us do have such codes. For example, engineers have an extensive code of ethics that covers a range of topics from public safety to competent performance. Part of their code of ethics reads:

Fundamental Canons: Engineers, in the fulfillment of their professional duties, shall:

1. Hold paramount the safety, health, and welfare of the public in the performance of their professional duties.

2. Perform services only in their area of competence.

3. Issue public statements only in an objective and truthful manner.

4. Act in professional matters for each employer or client as faithful agents or trustees.

5. Avoid improper solicitation of professional employment.[3]

There are also extensive rules of practice that indicate specific ways of implementing the fundamental canons. In addition, the engineers' code specifies certain professional obligations.

A suggested Code of Ethics for Policy Scientists indicates in part that:

1. A policy scientist should not work for a client whose goals and values, in the opinion of the policy scientist, contradict basic values of democracy and human rights.

2. Policy scientists should explicate assumptions and should present clear value-sensitivity analyses, so as further to increase the judgment opportunities for their clientele.[4]

Perhaps the worlds of engineers and policy scientists, if not simpler, are more explicit than ours. Somehow I have a hard time thinking of the people with whom I have been involved in developing models being willing to fully accept all the conditions of work effort embodied in the codes I have just discussed, although I am also sure we could be very comfortable with many of the elements of those codes. More importantly, however, we need to recognize that what we might call the modeling community, or modeling team, is not only limited to model builders and users. It includes a wide range of people in both the public and private sectors. These people, besides the builders, include the model initiators, the model evaluators, the "hands-on" model users, and the users of the model output. Of these, the model initiators, builders, evaluators, and "hands-on" model users might be either inside or outside the organization in which information generated from the model is being used.

It is my hope that all of the people involved in the modeling process behave ethically. But what, really, is ethical behavior? On a daily basis in our work, we see opportunities for unethical behavior, actual occurrences of unethical behavior, and lots of examples of behavior that is in a gray zone, seemingly ethical or correct for one person but not for another. We face many ethical dilemmas. Of one thing, however, I am fairly confident: ethical behavior is not related to one's competence.

As a user of models, let me share some examples with which I am personally familiar and for which the ethical considerations are not very clear. I will provide two categories of examples.

The first category relates to ethical dilemmas that might seem small in their consequences but are (1) important to the individual, and (2) cumulatively quite significant. I will provide two of these from the private sector.

The second category relates to ethical dilemmas of broad significance to both the individuals involved and the individuals affected. Due to the sensitive nature of these dilemmas, I will provide an example that has already been discussed in public: the decision not to adjust the 1980 Census count.

In the first category, the first case relates to a competent modeler who built a model to forecast the demand for a certain product. Due to the uncertainty inherent in this particular forecasting procedure, the result of the analysis was a point estimate that lay within a reasonably large range. The output of the model was to be supplied to an intermediate internal corporate customer who, in turn, would give it to another internal customer who was directly responsible for the sales of that product. Prior to giving his presentation on the model output, our modeler knew that the point estimate and the range of the model output was not within the "comfort zone" of the final user, who was responsible for the sale of the product. Our modeler was convinced that if he presented the range of model output, the ultimate customer would believe and act on only those data from the extreme range of the output. His assessment of the model's use was supported by considerable experience in dealing with this customer on this and other projects. He presented the range to the intermediate user, and then they jointly faced an ethical dilemma: should they present the range, knowing, in their own minds, that it would be misused and not represent the full knowledge available from the model? Or should they present only the point estimate, knowing that this was not all the knowledge available but that they stood a better chance of influencing the company's decision process in a way that they thought was correct? In this instance, they decided to present only the point estimate.

This example dealt only with people entirely within the firm. There are other examples of possible ethical dilemmas that involve people both inside and outside the firm. One way that models are built for us at General Motors (GM) is to hire outside contractors. There are many benefits in this process for us, including those that flow from the wide range of experience of the outside people. On the other hand, contractors do not completely understand the internal workings of our firm, so we need technically competent people from inside to serve as the liaison between outside modelers and inside model users. Thus, contractors selected to build models work with an internal analyst who "translates" the work of the contractor into input useful to our decision process. In this role, our internal analyst has the responsibility of understanding completely what the contractor is doing and understanding enough about GM to make some rather significant judgment calls mostly about "technical" assumptions in the model.

In the motor vehicle industry, so many factors come to bear on the decision process individuals use when they purchase vehicles, that there are always more factors than can be modeled. To address many of the factors, our internal analyst has to make assumptions. Each day he or she is faced with the necessity of deciding if the assumptions are correct, nearly correct, need more back-up work, or should be checked by someone else, and if they will make any difference in the model anyway. In addition, the analyst has to decide if he or she should share his or her uncertainties with his or her superiors, prematurely raising questions of validity of the model by listing all the technical problems he or she is running into without knowing their eventual impact on the model's validity.

Now, I am sure we can conjure up examples far more dramatic than these, but these are the day-to-day examples that often go unseen. Yet, their cumulative effect on the eventual use of models can be enormous.

Let me now move to the second category, which is a little more dramatic. In this case, the decision related to the question of whether the final count of the 1980 Census should be adjusted for the estimated number of people who were not counted.

There are several ways, many of which are orthogonal to each other, to look at that decision. When it comes to the entire Census itself, the perspectives are even more diverse.

Let me offer two perspectives within which I believe you can categorize many of the different points of view about the Census and which affect perspectives about the adjustment decision.

The first is that the Census is nothing more than a rather mundane, albeit costly, scientific event whose essential purpose is to provide the best estimate of the population of the United States.

The second perspective can be viewed as a much deeper clash among at least three systems: science, politics, and law. From this perspective, the story is the clash among three systems that have very different conceptions of what knowledge is and of how that knowledge can be produced and validated. I am simply going to touch on the ethical dilemma (as well as the frustration) faced by the statistician when involved with the law.

Consider yourself to be the Director of the Census Bureau. Now consider the following: The 1980 Census has not officially started, and a Federal Court judge renders a judgment that, in effect, orders you to prepare a plan (read model) for adjusting the 1980 Census. The judge's decision was predicated on a mixture of (1) facts: there had been sizable differential undercounts in prior Censuses; and (2) assumptions; several experts (mostly statisticians — some actually from the Census Bureau) had indicated at the trial that a differential undercount was to be expected in 1980 because the social phenomena associated with those persons who were difficult to count had not changed from 1970. In short, the court anticipated or presumed that there would be a sizable undercount, which can be described as the determination, by means other than a direct count, of the number of people who have been missed by the direct Census count. It requires, of course, an estimate of what the count should be. The foundation of this estimate lies in the assumption that the population of the country at a given point in time is equal to the population at some earlier time plus all births and immigrations during that time interval minus all deaths and emigrations during the same time. In essence, it is a fairly complex estimate of how many people are missed during the enumeration.

Now, as the end of the enumeration period approaches, it becomes clear to you that, according to all the ways by which you have estimated the undercount in the past, the preliminary tabulations tell you that whatever the undercount might be, it has a high likelihood of being both lower than and different in its nature from the undercurrents in the past. One particular reason stands out. Due to some of your coverage improvement programs, you have counted a lot of people who were not documented citizens of the United States, and since they are not in the denominator of the expected count but are in the numerator of the actual count, you have, like it or not, reduced the "measured" undercount dramatically. You find yourself in a position of knowing that if there is an undercount, it is not measurable using the accepted procedures you have traditionally used, which are also the very same procedures that the plaintiffs had used to convince the judge that there would be another undercount. Armed with this new information, you inform the judge of these new findings, believing, as a scientist, that if the basic assumptions that supported his earlier decision were changed, he would reconsider his decision. The judge's response makes it clear to you how wide the gap can be between science and the law.

The judge, in essence, says, "I heard your comments, but even if there eventually is no undercount, you can report that fact at the appropriate time". The judge also tells you that you have been non-responsive insofar as you have concluded, based on this new evidence, that no statistically defensible method exists for adjusting the Census count as it will likely turn out. Quite simply and directly, the judge says, "I said adjust, and left you only with the question of which method to use".[5]

As you prepare to deal with this order, another judge in a different jurisdiction is ruling on a very similar case brought by another plaintiff. This judge, after making some comments about how disputes like this should not be resolved by the courts and about the need to keep a national perspective on the count and not be swayed by local considerations, closes his opinion by saying, "So while the Bureau may not fulfill its duty arbitrarily, capriciously, or fraudulently, it nevertheless should be afforded wide latitude to select its methods and techniques, and by those means to arrive at its own result". [6]

Your first reaction could very well be, "Why me? One judge tells me I'm being unresponsive because I honestly come forward and tell him that the basis upon which he ordered me to do something no longer exists. The other judge says that, as long as I'm neither arbitrary nor capricious in the conduct of my activity, I should do what I believe to be best".

Now, just about the time you start to wonder why you ever took this job, your attorney, the Justice Department, advises you that, like it or not, the judge is the judge. You must tell him how you're going to adjust the Census, even if you do not know of any statistically defensible way of doing it.

Well, here you are facing a very interesting ethical question. If you say,

"I can't do it", you're violating the law. And if you do that, they'll bring other people in, until someone agrees to do it. If you do select a method, you are giving credence to something that you strongly believe doesn't warrant the acknowledgement, however conditioned, that it was chosen by the Census Bureau to adjust the Census count.

What did the actual Census Director do? After much anguish and gnashing of teeth, he in essence told the judge, "Since any of the methods (read models) for estimating the extent of undercount from which we could derive an estimate would not be statistically defensible, we will identify two and provide the court with the strengths and weaknesses of each and let the court choose from among them. One would be less technically deficient than the other, while the other would be faster and easier to implement. As it relates to how we would distribute the estimated undercount across the nation's legal jurisdictions, we would use the least expensive of methods, since none of them, in this instance, would be defensible".

Well, we might feel that the Director did the "ethical" thing (or evaded the ethical issue by sandbagging the judge), but needless to say, the judge was not pleased. Nor, I might add, was he pleased when his decision, as well as every other decision demanding adjustment on the basis of the grounds he was proposing, was overturned on appeal.

Given that the types of problems I have identified permeate many of our modeling activities, what type of code of ethics should we have for the modeling community? What are the options? First, there is the traditional professional code of ethics, such as those endorsed by attorneys, physicians, or engineers. However, these groups of professionals are often relatively homogeneous, in that they are educated in a common manner, licensed in a uniform manner, and have defined professional oversight bodies that can levy sanctions for ethical violations.

We as model builders and users come from exceedingly diverse educational and professional backgrounds. It would be very difficult to enforce a code of ethics within our ill-defined community, probably even harder than it would be to decide what the code ought to include. We have no professional society or organization encompassing all relevant people. In addition, what could possibly be the punishment for someone who violated the code? We have no licenses that could be revoked. It is hard even to imagine that a code, by itself, would be successful in producing widespread ethical behavior. We already have many general social codes that encourage ethical behavior.

Let us think about an alternative way of enhancing ethical behavior among all members of the modeling community. It has its roots in another law, this time not the recent Barabba's Law, but the much older law: "Caveat Emptor" — Let the Buyer Beware!" This approach can foster better development and use of models by having a set of questions a model builder should be prepared to answer and a model user should ask and expect to have answered. Model builders and users would agree to the content of the questions. This is an open way to educate all participants in the modeling process about models and their limitations and should result in better models and better use of them.

This is in the best interest of the entire modeling community, because if models are not correctly used or not used at all, then modeling will soon fall into disfavor (or as some might say, further disfavor). Communication between the model builder and user about the nature of the model is required. Communication about its performance capabilities is crucial. Greenberger and his colleagues have emphasized the problems that can occur because of poor communication. They said, "An essential part of the modeling process is the translation of the policymaker's perceived concerns into researchable questions. If the policymaker is inarticulate, aloof, or ignorant of the capabilities and limitations of modeling — or if the modeler lacks empathy with the policymaker — the enterprise may go astray before it has fairly begun". [7]

Let me mention some of the types of questions that a model user would want to ask a model builder. These questions should be agreed upon by both, and all other relevant parties should also have access to the answers. These parties are those I mentioned above and might include the model initiator, the hands-on user, the evaluator, or the final user. Several of the following questions were originally posed by Barbara Richardson and her colleagues at the University of Michigan in 1979.[8]

1. How well does the model perform?
2. Has the model been analyzed by someone other than the model authors?
3. Is documentation adequate for the users' needs?
4. What assumptions and data were used in producing model output?
5. Why is the selected model appropriate to use in a given application?
6. Will the model be run directly and specifically for the present purpose?
7. What is the accuracy of the model output?
8. Does the structure of the model resemble the system being modeled?
9. Is the model appropriately sensitive to the inputs being varied?

In addition, the model builder could introspectively ask the following questions when preparing the model:

1. Do I understand the problem?
2. Are the assumptions reasonable?
3. Is the method I have selected the best? Are there other possible ways to do it?
4. Can I get this done within the time frame available?
5. Are the data I am using to build and run the model correct?

But these questions need a framework within which we can ensure that their answers will have real impact. A first step toward providing a framework or environment for ethical behavior to evolve is to develop a consensus within the modeling community over the types of questions that users should ask of models and modelers.

This, by the way, is not much different than the approach I have used to encourage improvements in the use of market information. The approach we have been using has been described as the Inquiry Center.

The concept of the Inquiry Center, again, is not new news. It was proposed by C. West Churchman in his 1971 book, *The Design of Inquiring Systems*.[9] In it he proposes the development of inquiring systems that support the decision-making process. One of the foundations of the Inquiry Center concept is the acceptance of different forms of inquiry.

Let me reinforce this notion of the need to be capable of dealing with different forms of inquiry by quoting from Ian Mitroff's very interesting book, the *Subjective Side of Science*, his characterization of four of Churchman's philosophical inquiry systems.

Leibnizian Inquiry Systems

Leibnizian Inquiry Systems are the archetype of formal-deductive systems...emphasiz(ing) the purely formal, the mathematical, the logical, and the rational aspects of human thought. They represent the side of scientific inquiry that has always been interested in the construction and exploration of purely theoretical models. The essential characteristics...can be captured as follows: Leibnizian Inquiry Systems start from a set of (1) elementary, primitive (i.e., undefined) explanatory variables or primitive truths, and from these, they attempt to build up through (2) formal operations or transformations, increasingly more general or universal (3) formal propositions or truth nets. For all practical purposes these formal truth nets (linkages of propositions) may be regarded as the information output, or better yet, as the information content of a system.

The strengths...are the strengths that characterize all formal systems: consistency, rigor, logical coherence, precision, little or no ambiguity in the use of terms, conditions of proof, and so on. Their weaknesses are the weaknesses that beset all formal systems; for all their emphasis on logic, precision, and rigor, Leibnizian Inquiry Systems are often extremely hard put to defend...why they chose to solve a particular problem and why they chose to represent it in the manner used.

Lockean Inquiry Systems

Lockean Inquiry Systems are the archetypes of experimental, inductive, consensual systems...emphasiz(ing) the purely sensory, empirical aspects of human knowledge. Where Leibnizian Inquiry Systems build up increasing more general — formal — proposition truth nets from elementary — primitive — analytical truths, Lockean Inquiry Systems build up increasingly more universal inductive generalizations (fact nets) from the elementary sensory data of raw experience. While in the Leibnizian Inquiry System it is a set of formal operations that transforms the primitive elements into elements of information, in the Lockean inquirer it is the function of human judgment which accomplishes the transformation from raw data to factual information.

The strength...lies in their ability to sweep in rich sources of experimental data. In general, the sources are so rich that they literally overwhelm the current analytical capabilities of most Leibnizian systems. The weaknesses...are those that beset all empirical systems. Although experience is undoubtedly a rich source of knowledge, it can also be extremely fallible and misleading. Further, the simple sensations, facts, or observables of the empiricists have always, on deeper analysis, proved to be exceedingly complex and further divisible into other entries themselves thought to be indivisible or simple, *ad infinitum*.

Kantian Inquiry Systems

Kantian Inquiry Systems are the archetype of synthetic multimodel systems...emphasiz(ing) both the formal and the experimental — the integrative — aspects of human thought. They are synthetic in the sense that they attempt to reconcile (synthesize) the demands of the philosophic rationalism (Leibnizian IS) and that of the philosophic empiricism (Lockean IS). They are multimodel in the sense that, where Leibnizian and Lockean

Inquiry Systems usually build only one formal model or only one inductive generalization, Kantian inquirers produce at least two alternate models, either of which will equally fit the data or explain the primitive explanatory variables.

The strength of the Kantian Inquiry System is that it counters the weaknesses of both Leibnizian and Lockean Inquiry Systems. The weaknesses are: (1) there is no guarantee that the multiple models will include the right model; (2) there is the danger that the decision maker will be more overwhelmed than aided by the multitude of models; and (3) Kantian Inquiry Systems are more costly to operate — with the addition of each model, the cost and time of the system goes up disproportionately.

Hegelian or Dialectical Inquiry Systems

Hegelian or Dialectical Inquiry Systems are the archetype of conflictual synthetic systems...emphasiz(ing) the antagonistic and the antithetical, the conflictual aspects of human thought. An Hegelian inquirer is designed to present the strongest possible debate on any issue. In a Kantian Inquiry System the alternate submodels are not necessarily antagonistic. They may in fact be highly complementary. There may be a great deal of overlap between them...Is a Hegelian inquirer, the overlap (or in set theoretic terms, the intersection) is zero. In a Hegelian inquirer, the submodels (of which there are at least two) are in complete opposition on almost any and every point. As anyone familiar with Hegelian thought knows, the conflict runs so deep that the two opposing points of view are the deadly enemies of one another.

The strengths of an Hegelian inquirer are: (1) the decision maker is actively involved in the information creation process; unlike some of the other inquirers, the user of the information is no longer a passive receptacle for the end-product but an active creator of what he will use; (2) unlike most scientific information systems that speak in the dry language of facts and symbols, Hegelian Inquiry Systems speak in the more active language of drama. They use conflict both as an intellectual and as a dramatic device to induce the decision maker to take an active interest in the system. The weakness of Hegelian Inquiry Systems is that they may create conflict when it is not there or appropriate. Further, not all decision makers can either tolerate conflict or learn from it optimally. In addition they are costly and time consuming because they involve the creation and use of at least two equally credible theories or experts.[10]

I have found that the Hegelian model seems more appropriate to my style of inquiry, particularly in today's complex world, where often the challenge for a manager in decision making is not to seek the right decision. Rather, it is to manage the process in a way that increases the chances of choosing the best decision among the available alternatives, given all the circumstances at that time.

Indeed, in the case of determining whether or not to adjust the 1980 Census, we used a Hegelian model, sometimes referred to as Strategic Assumption Surfacing and Testing (SAST), supported by Ian Mitroff and Dick Mason. In fact, there is a not-very-well-read book entitled, *The 1980 Census: Policy Making Amid Turbulence,* by Ian, Dick, and myself, which, in great detail, tells probably more than anyone needs to know about the subject.[11] Actually, the process served us quite well. For example, when we explained to the courts that we used a process which was designed to reveal both sides of an issue, that we published the results of our findings, and that we allowed interested parties to challenge our own set of assumptions, it certainly reinforced our point that we were being neither capricious nor arbitrary in developing the decision on whether or not to adjust.

However, whatever model you choose, or all of them for that matter, the Inquiry Center must provide the appropriate tools, expertise, and innovative momentum to facilitate the decision process.

The Inquiry Center must allow for an organized collection of information to be integrated into the decision-making process of the organization. That is, the system must be adaptable to the environment in which it will operate and be considered an appropriate inquiry tool of the decision maker.

The new inquiring systems must be multidimensional. They must integrate the logic of decision making with the energy and the imagination of those who will affect, or be affected by, the outcomes.

The first of these dimensions is the logic/analytical dimension. We are accustomed to decision making as an extension of the logic/analytical process. At the heart of this approach is the "process of separating or breaking up a whole into its parts so as to find out their nature, proportions, functions, relationships, and so on". In Churchman's terms, this is primarily the world of the Leibnizian inquirer.

The second dimension is the energy/organizational (collaboration) dimension. Decision making and implementation in virtually any organization is a collective process. It is shaped by the factors and dynamics that underlie human behavior in social settings. Sharing of that information from various perspectives is a start. Here, both the Kantian and the Lockean inquirer would be most comfortable.

The third dimension is formed by what might be called "imagination" or "creativity". People with access to diverse experiences and viewpoints can be the key to generating the best possible range of alternatives to solve an important, complex problem. Here the Hegelian inquirer is most likely to make a contribution.

What our experience to date tells us is that what the Inquiry Center needs most of all is a conducive environment. By conducive environment, I mean explicit sponsorship by senior management, recognition for the use of tools, consideration of appropriate representative participation, skillful facilitation, and attention to implementation planning for successful closure. This environment could be quite costly, both in terms of dollars and in terms of the dramatic change in operating style that may occur.

However, for some decisions that face businesses and society today, the cost of creating this environment is worth the effort if it ensures that the decision-making team has the time and environment necessary for careful consideration and the opportunity for creative breakthroughs.

In essence, then, the Inquiry Center is the "home" for the tools within a three-dimensional construct. It is an arena in which all the functions of a decision-support system are bundled during the decision process. It is set up to enhance the integration among the three dimensions.

Currently in operation at General Motors are several Inquiry Centers, though none of them are called Inquiry Centers. In an effort not to slow the concept down, we have chosen not to make it sound as if we are creating something new — and indeed, we are not. One of these includes members of the Market Research and Planning Department and the Design Staff, and is attempting to help the market research community learn how to present what we know about the market to this very artistic and creative community in ways that will help them incorporate that knowledge into their vehicle concepts. Another is with the Advanced Engineering Community (the other side of the world) where we are attempting to resolve similar communication problems, but in this instance, we are including the engineers in the design and collection phase of market research activities. With both of these projects, we are learning how we can make significant progress toward reaching a common understanding of what our customers really want in a vehicle and translating those desires into our new cars and trucks.

To my mind, the same type of thing can be done to facilitate the more efficient, effective, and ethical use of models. With all members of the modeling team participating in an Inquiry Center, with the goal of providing input for quality decisions in the firm, great strides can be made toward achieving a common understanding of the problem, understanding it within the constraints of a model, and presenting the results as honestly and meaningfully as possible. We do not want our analysts to have to decide if a truthful or untruthful response will yield a greater benefit to our firm. A user should understand what a "range" means so that model output can be shared.

In summary, then, let me state my main ideas on ethics in modeling for your consideration. First, I think that it is very important for us to address ethical behavior in the modeling community. There are just too many opportunities for opportunists.

Second, I do not think that we modelers, as a professional group, have enough cohesiveness, common background, and experience (partially because a "model" is such an ill-defined construct) to develop and enforce a code of ethics that will be meaningful or functional.

Third, I believe that in order for our profession to continue to function effectively, and indeed, to survive, we need to make ourselves more useful and effective by ensuring that the users of our models know what it is that they are buying. Let the buyers beware and by doing so, be better served!

The key to the effective use of models is the knowledge of what the model can and can't do. It is incumbent upon model users to ask a whole range of questions in order to understand as much as possible about the model to be used. In order to facilitate such understanding, I propose that working environments similar in nature to the concept of an Inquiry Center be established. These will allow the open and continued exchange of information among all members of the modeling team, allowing consensus to be reached and, in the process, the ethical use of models to be encouraged and enhanced.

In this way, we will not be required to overpromise on our models to ensure we get access to required resources. But what about Barabba's Law: "Never say the model says?" Well, if we do not have to oversell, then, alas, we do not need Barabba's law!

NOTES AND REFERENCES

1. Blitzer, H.L., "Commentary", *Marketing Science*, **6**(2), 175, Spring 1987.
2. Greenberger, M., Crenson, M.A. and Crissey, B.L., *Models in the Policy Process*, New York: Russell Sage Foundation, 323, 1976.
3. National Society of Professional Engineers. *Code of Ethics for Engineers*. NSPE Publication No. 1102 as revised, January 1985.
4. Quade, E. S. *Analysis for Public Decisions*. Second Edition. New York: North Holland, 348, 1982.
5. Klutznick *et al.* v. Young *et al.* U.S. Supreme Court, 21, October 1980.
6. Green *et al.* v. Klutznick *et al.*, no. 80-3142 US, 1980.
7. Greenberger *et al.*, *op. cit.*, 325.

8. Richardson, B.C., Joscelyn, K.B. and Saalberg, J.H., *Limitations on the Use of Mathematical Models in Transportation Policy Analysis*, Ann Arbor, MI: UMI Research Press, 11–13, 1979.
9. Churchman, C.W., *The Design of Inquiring Systems: Basic Concepts of Systems and Organization*, New York: Basic Books, 1971.
10. Mitroff, I.I., *The Subjective Side of Science, A Philosophical Inquiry Into the Psychology of the Apollo Moon Scientist*, Amsterdam: Elsevier, 1974.
11. Mitroff, I.I., Mason, R. O. and Barabba, V. P., *The 1980 Census: Policy Making Amid Turbulence*, Lexington, MA: Lexington Press, 1982.

Chapter 4

What are the Ethical Responsibilities of Model Builders?

Part 1. From Model Building to Risk Management: Evolving Standards of Professional Responsibility

N. Phillip Ross and Suzanne Harris

INTRODUCTION

Decision makers may not know much about mathematical models, but they know what they like: algorithms that make intuitive sense, plenty of predictive power, and narrow margins for error.

Models of complex processes seldom oblige. Statistical analysts appreciate the many ways in which models can misrepresent reality. As more of us make the effort to educate users in the correct interpretation of model results, we are evolving new standards of professional responsibility in which user education is a requirement.

Underlying the new standards is a growing acceptance of risk management, an approach to decision making that explicitly takes uncertainties into account. This trend has important implications both for model builders and for model users.

WHAT IS A MODEL BUILDER'S RESPONSIBILITY?

A model builder has a professional responsibility to explain not only the strengths of a particular model but its inadequacies — to point out how the model may fail to depict an underlying process correctly. The explanation may not be well received. Most users prefer "is" to "might be" and want a model to simplify a decision, not add to its complexity.

Therefore, educating users takes time. If a model builder's work is backlogged, and it usually is, management may regard user education as a diversion. Yet in a regulatory agency such as the Environmental Protection Agency (EPA), user education can be critical to the correct application of models in complex areas such as carcinogenic risk or global climate change. Model builders must do more than turn over a model and say, "Here is the answer." They need to familiarize themselves with the decision context and help users understand what questions can and cannot be answered with the aid of a model.

Users are often less analytically sophisticated than they seem. In the extreme, some may regard a model as a black box whose outputs are assumed to be correct if the analyst signed off on them. (We hope that these users are in a shrinking minority.) More typically, users understand the model's basic mathematical relationships but may need help in comprehending the underlying uncertainties.

Some users need to be taught to distinguish deterministic models from probabilistic models and to be cautioned against excessive reliance on a model as a depiction of future reality. Other users need a short course in stochastic error, expressed in terms that a non-statistician can easily grasp. For example, in the (unidentifiable) quotation below, *The Wall Street Journal* successfully expresses a statistical concept in terms that can be understood by a non-technical audience:

> For each poll, the odds are 19 out of 20 that if pollsters had sought to survey every household in the U.S. using the same questionnaire, the findings would differ from these poll results by no more than 2½ percentage points in either direction. The margin of error for subgroups would be larger. For example, for working women, the margin of error would be five percentage points in either direction. In addition, in any survey, there is always the chance that other factors such as question wording could introduce errors into the findings.[1]

Almost all users need to be reminded that a model can truthfully represent the data at hand and still be wrong, wrong, wrong. As we know, model builders can be thwarted by problems ranging from sloppy data collection to missing variables. If the processes that generate errors are hidden from the model builder, probabilities cannot be assigned; still, decision makers need to be aware that such errors may exist.

??? x 10⁻? AND OTHER MYSTERIES

A model can misrepresent the world in many ways. Problems commonly arise when a model's inputs originated as outputs from another model. Uncertainties in the first model are thereby multiplied by uncertainties in the second. Once users grasp this concept of cascading risk, they understand that the final result of this chain of output-input-output falls within a wide margin of error.

Many users have found this a hard concept. Until fairly recently at EPA, for example, the only influential data in many regulatory decisions concerning environmental health regulations were point estimates of risk, e.g., "123.4 x 10^{-6} excess cancer deaths per year" attributable to emissions of a certain chemical. Certainly, we need to have some way of deciding when to act. However, epidemiological data relating the incidence of cancer to environmental factors is very scarce. Most of our data reflect only occupational exposures. Consequently, we use experimental bioassay data and mathematical models that extrapolate from experimental data to estimate the effects of exposure of humans.

Our understanding of the biological mechanism of carcinogenesis is far from perfect. Therefore, we tend to choose mathematical models that we believe give us the widest margin of safety in estimating risk. We observe the incidence of cancer in animals receiving high-level dosages of the chemical that concerns us. We take those dose-response points and fit curves to them. We do not know what the low-dose points would be because the incidence of cancer at low doses is so low that it would require doing enormous experiments with hundreds of thousands of animals, which we know we cannot do.

Therefore, we give a manageable number of animals high doses of suspected carcinogens. Thus, the response points are all at a level of exposure that never occurs in a natural situation, we fit a mathematical function to those points and extend it to the low points. The trouble is that several mathematical models will fit those upper points with different tail distributions.

Which model is the right one? We do not know, so we choose conservatively, possibly overestimating risk to humans but probably not *under*estimating risk.

We go on to make further choices. We select a model to extrapolate the results from laboratory animals to humans. Then we select a model to depict human exposure to the chemical of interest. Exposure ("fate and transport") models depend upon ambient monitoring data, and many of these data are obtained at sites far removed from intensive human exposure —— airports, for example. How different might the results be if monitoring had taken place in the central business district?

Thus, we have not only the uncertainty associated with experimental error — sampling error, design error, all the other sources of error in a normal controlled experiment — we also have random error, and in addition, by asking what would happen at low doses, we enter into a new realm of error, which is model selection. It is very difficult to quantify that error. As we indicated, scientists tend to select conservative models to err on the side of protecting public health. At the same time, they conduct pharmacokinetic experiments in the field to map into the actual biological processes.

If the model builder is fortunate enough to get involved at the experimental design stage, the most difficult type of error — bias, or systematic error — can perhaps be minimized. Later on, it simply must be acknowledged that a model could misrepresent the real world in ways that are completely unknown and, therefore, cannot be explained using a probability distribution.

RISK MANAGEMENT IS GAINING ACCEPTANCE

EPA is currently integrating quantitative risk assessment into an overall decision context known as risk management. In risk management, the quantitative analysis takes its appropriate place in the decision process as input to the decision, but is not the driver of that decision. Risk management recognizes that any single number suggests an unrealistic level of precision and accuracy. In the same vein, statisticians are moving away from arbitrary use of the 0.05 confidence interval to express statistical significance. Instead, we are allowing the researcher to define significance in the context of the project. It is becoming more common to see: What is the probability of a Type I error? A Type II error?

At EPA, global climate change modeling is a particularly visible area in which modelers and decision makers have grappled with multiple sources of potential error. Attempts to predict how the climate will change use a complicated array of models and submodels operating on very scarce empirical data. As the economic impacts of regulatory action are likely to be severe, it has been especially difficult to act before more of the uncertainties are resolved. We understand that we are collecting more data on temperature and sea level, which may help to resolve much of the current controversy.

Uncertainty is only one of many factors that can complicate model results from the user's standpoint. There are also the many potential sources of bias, including experimenter effect, halo effect, and other phenomena.

Users can disregard these complications; in fact, many do. However, a responsible modeler will educate the user to understand bias and to look for it in a model, even if this reduces the user's confidence in the model's results.

BEWARE THE MODELER WHO LEAVES NO TRACE

Another of the modeler's responsibilities is documentation, the absence of which has soured many projects. In the world of science, results published in a journal can be tested by duplicating the study to see if the results are the same. What we often see in modeling are situations in which no one can duplicate what the model builder did because no one completely understands how the model works.

Missing documentation is a frequent complaint about contractor modeling projects. Contractors often save documentation until last or allow it to fall between the cracks if funding runs short. A cynical contractor may omit documentation in order to remain indispensable to the project.

If a regulatory agency is to command respect, decisions must be consistent. Agencies use models on a regular basis to promote consistency by producing results that users perceive as logical. When different users run a model with different input variables, they must be able to compare notes and understand how the model generated their respective results.

This seems self-evident, yet a surprising number of model builders fail to provide users with complete documentation in non-technical language. The documentation may omit a component of the algorithm, or the modeler may use transformations that can be understood only through tedious scrutiny of the computer code. As a result of such omissions, users may be unable to reconcile the model's results with their intuitive understanding of the model's logic. This is likely to destroy their confidence in the model — and may also destroy confidence in decisions made using the model's results.

PUTTING PROFESSIONAL RESPONSIBILITY INTO ACTION

Aware that most users are not trained as statisticians or economists and may not necessarily be experienced in applying model results, a responsible modeler will try to shed light on the process. As we indicated earlier, no one pretends that user education is easy. The modeler needs to frame the explanation in terms familiar to the user and must resist the temptation to use overly technical terms. Processes that seem obvious to the modeler may need to be explained in detail.

The modeler should take the initiative; many users will hesitate to ask questions because they do not want to admit to being confused. Early in the model-building process, modelers should elicit from users a clear explanation of the way in which the model will be used in decision making. Later on, this understanding can help the modeler explain aspects of the model that are most relevant to the user.

Modelers should begin by laying out key assumptions about the process that is modeled. Technical details are probably less important to the user than major ways in which the model's results may fail to depict accurately the observed process. This type of uncertainty should be discussed at length.

Finally, the user should be encouraged to ask questions. The modeler's ultimate goal should be a model that can become a permanent fixture in decision making. Only if users thoroughly understand a model and its limitations will they feel confident enough to institutionalize its use as a decision-support tool.

Chapter 4

Part 2. On Model Building

John D.C. Little

INTRODUCTION

Models are central to scientific thinking and essential to applying scientific knowledge to practical problems. In this chapter we explore four aspects of model building: (1) the meaning of the word — examination of its usage reveals multiple perspectives; (2) models for science vs. models for problem-solving — different goals lead to distinctively different approaches for model construction; (3) the process of building models — where do they come from; how are they built; and (4) social responsibility — a look at potential sources of ethical dilemmas.

WHAT DO WE MEAN BY MODELS?

The word, model, turns up frequently both in conversation and in writing and, like any popular word, means different things to different people. Going back a century or two, physical scientists believed they were discovering the laws of nature. We find Newton's laws of motion, Kepler's laws for describing the movement of the planets and Snell's law for the refraction of light, to name a few. However, so many of these laws began to be repealed, or at least amended (Newton's law, for example, by Einstein's relativity theory) that a more conservative and tentative word came into use — models. Later, when the social scientists started to quantify their world, they did not find relationships among their variables that were as precise and complete as in the physical world. So they embraced the word quite naturally. Model captures an important texture of tentativeness and incompleteness that we often want when describing our knowledge. Biologists, too, have adopted the term but in a rather different way. By a model they mean a biological system that can be used as an analog for another biological system. Thus, a genetically altered mouse might be a model of a human in terms of responding to a particular type of cancer.

I find that my own use of the word depends on the person to whom I am talking. With fellow model builders, I mean something fairly precise; when talking to others — perhaps managers or dinner guests or newspaper reporters — either I do not use the word at all or I use it rather diffusely. My informal definition among model builders is that a model is a mathematical description of how something works. Good examples are: $F = ma$, distance = rate x time, and the economic lot size model. When I use the term among such people, it connotes structure, mechanism, functional relationships among variables, and inputs determining outputs.

However, there are nuances and quality differences. Here are a few:

Statistical Models

Is linear regression a model? Obviously, yes. Yet, linear regressions represent such a broad and flexible class of models that they contain within themselves less *a priori* structure and less information than, say, most physics or OR/MS models like $F = ma$ or the economic lot size model. Each of the latter makes a strong structural statement about a specific phenomenon. A goal of the statistician is to provide estimation procedures and probabilistic statements that apply as generally as possible. Therefore, statistical models tend *not* to say much about phenomena until particularized to a given situation by application. Often the convenience of data analysis afforded by standard statistical models leads the user away from building deep structure for problems to less powerful models than might be desirable.

Decision Analysis

Is decision analysis a model? Well, wrong question. Analysis does not equal model. Decision analysis provides a logical structure for the decision making process, i.e., it specifies a series of problem-solving steps that a decision maker can take. A decision analysis application may also contain one or more explicit or implicit models about how the world works: events are defined and assigned probabilities, the utility of the decision maker is related to possible outcomes, and outcomes are related to alternatives.

Procedures and Algorithms

Is the simplex algorithm a model? Is the exponential smoothing of a time series a model? No, I say, although models are lurking in those activities. Procedures and algorithms certainly have inputs and outputs just as most models do. However, I think our language is richer and more useful if we make a distinction between model and procedure. Thus in the linear program

max $\quad\quad\quad\quad z$
subject to

$$z = cx$$
$$Ax = b$$
$$x \geq 0$$

the last three lines constitute the model. Our normative aspirations are expressed in the max and the simplex method is a way of finding a solution, but these are separate from the model itself. Similarly, I would tend to view exponential smoothing as a computational procedure. Distinct from the calculation is an underlying model of a process, represented by certain recursive equations and probabilistic assumptions.

Social Science Theories

In social science the term model appears all the time when there is no mathematics anywhere in sight. Model here means theory: a system of related concepts to describe an idea or phenomenon. Quite frequently, I use model in this way too, particularly with people who are not mathematical model builders. The word helps convey the tentative and incomplete nature of the theory.

These then are some niceties in the way the word is employed. Many people in OR/MS would likely share these views, but other disciplines may have somewhat different usages.

MODEL BUILDING FOR SCIENCE

Building models for science differs from building models for OR/MS or engineering. The differences lie in the criteria employed, both for choosing what to model in the first place and for judging the model when it is finished. Furthermore, the process of building the model changes.

Science is concerned with describing the universe with fidelity and parsimony. These fundamental criteria tend to determine which work survives to be recapitulated in the text books of the next generation, although scientists care about other qualities as well. They respond to elegance, beauty, and surprise. Problem selection is also affected by personal motivations that go unrecorded in scientific journals: salary, promotion, tenure, and Nobel prizes.

Scientists have developed a variety of tests for assessing fidelity: they think up and enumerate threats to validity, try to falsify a result, develop alternative hypotheses and devise critical experiments or observations that will discriminate among them. They make predictions and compare these to observed outcomes. Such devices help convince scientists and their peers of the robustness and generality of their discovered knowledge.

MODEL BUILDING FOR PROBLEM SOLVING

Most of us in OR/MS, policy analysis and the like are trying to help organizations make improvements in the world, or at least in their corner of it. Having a goal tends to change and clarify the model-building process. Interestingly, I can think of many more how-to-do-it lists for the problem-solving side of model building than for purely scientific work.

SYSTEMS ANALYSIS

1. Define Problem

    Scope
    Objectives
    Measures of effectiveness
    Constraints
    Control variables

2. Develop Alternatives

    Identify subproblems
    Determine alternate ways of
      solving subproblems
    Synthesize alternative
      overall solutions

3. Evaluate Alternatives

    Formulate model of system
    Calibrate model from data
    Evaluate alternatives
    Select best

4. Satisfied?

    No. Recycle
    Yes. Quit

Fig. 1.

The first difference is that in problem solving we presuppose a client or customer. This might be a manager, an organization, or possibly society as a whole. The model-builder is a consultant and model building is imbedded in a larger consulting process. The principal goal becomes to improve the client's welfare (although we should keep in mind potential conflicts of interest relating to the model-builder's welfare). New criteria for model quality come into play. Fidelity and parsimony are welcome but not at the expense of relevance. It may be better to be vaguely right than precisely wrong.

People have devised a variety of how-to-do-it problem solving paradigms that relate to model-building. We shall briefly describe three: (1) systems analysis, (2) the phases of OR, as presented in an early OR text,[1] and (3) one of my favorites, Urban's "Building Models for Decision Makers".[2] The last is interesting because it explicitly recognizes the role of model building within the consulting process itself.

Figure 1 shows a step by step recipe for systems analysis. It consists of the stages: defining the problem, developing alternatives, and evaluating the alternatives. The process anticipates multiple iterations through some or all of these steps, since system analysts do not conceive a model in one fell swoop but cycle around, developing ideas and fitting pieces together.

Paradigms like this are useful. They provide check lists. People who have been working on problems develop heuristics, constructs, and chunks of useful knowledge for solving them. A programmed list helps jogs such materials out of memory and get them used in an organized way. Key words direct one's thinking along certain pathways that need to be examined.

Although none of us treat such prescriptions rigidly, they contain considerable wisdom. It is fair to say that Fig. 1 presents, in terse form, some really important ideas about the structure of this kind of problem solving: the notion of defining the problem — that it must have finite and defined scope and that you need not only broad objectives but also narrow measures of effectiveness with which to compare alternatives. The world is constrained and you should explicitly examine those constraints (perhaps with a view to modifying some of them). Obviously you have to identify the variables and other actions you can control. There is little that is more fundamental than the idea of developing alternatives or of breaking a problem into smaller problems and trying to synthesize solutions from manageable pieces. The notion of evaluation is critical, as is the role of models and data in doing this. Finally, we have the necessity of choice.

PHASES OF OR

Churchman, Ackoff and Arnoff (1957)

1. Formulating the problem.

2. Constructing a mathematical model
   to represent the system under
   study.

3. Deriving a solution from the model.

4. Testing the model and the solution
   derived from it.

5. Establishing controls over the
   solution.

6. Putting the solution to work:
   implementation.

Fig. 2.

Churchman, Ackoff and Arnoff's[3] phases of OR in Fig. 2 is shorter and less rich, but introduces several ideas not in the systems analysis description. These include testing the model and an explicit concern for implementation, as well as the concept of "controlling the solution". By this phrase the authors mean that the environment of the problem may change and render the proposed solution invalid. Thus, the solution needs to be controlled in the sense of quality control for an industrial process.

Urban[4] has gone a step further, as shown in Fig. 3, drawing on organizational development ideas, particularly those of Kolb and Frohman.[5] He points out that we do not approach a practical problem with a blank sheet of paper, but in fact arrive with priors about what the issues are and how we might model them. Furthermore, our success in actually effecting change depends critically on how we enter the organization. Do we come at the request of the president? Or are we trying to sell ourselves in at the bottom and work up through several organizational layers to the level at which our work would be used? Do we come through a gatekeeper, for example, through marketing research to reach marketing management? These are important questions and, if we are really going to help the organization, they are critical strategic issues for us as model builders and prospective change agents.

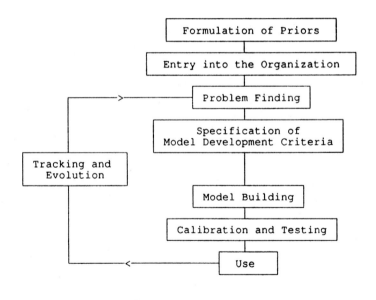

BUILDING MODELS FOR DECISION MAKERS

Urban (1974)

Fig. 3.

Once we are in the organization, there comes the question of finding and defining the problem. Ordinarily, problem finding works by uncovering a discrepancy between where the organization is on some key dimension and where it would like to be. Following this stage, Urban introduces another step not mentioned in most standard prescriptions: the specification of model development criteria. This means establishing with the client rules by which the model results will be judged. It is a key step that is often omitted or insufficiently treated in consulting practice. The result is frequently misunderstanding at a later date. The client remembers the sales pitch and the consultant notes the vagueness of the actual contract. Discussion is needed early in the engagement to produce a mutually understood document that clarifies expectations. The handling of this step has much to do with the client's satisfaction or dissatisfaction with the project.

Following this is the model-building stage itself. Urban's prescriptions include calibration, testing and, finally, model use. Then in situations where the models will have ongoing application, there is tracking of predicted vs. actual outcomes and evolutionary improvement of the model. Such a feedback and incremental improvement process is common in decision support applications.[6]

Problem solving prescriptions like these mean the most to people who have tried, at least to some extent, to perform the process. To those who have not, the paradigms seem relatively opaque and meaningless. I find that undergraduate students are likely to see such recipes as so much hot air, but after a summer job trying to solve practical problems, they relate to the ideas much more. A really experienced person is also likely to find the prescriptions superficial because the main principles have long since been internalized and second-order subtleties have become salient.

A final point is familiar to all model builders but totally foreign to almost all clients: A model should be imperfect. It should fit the task at hand, i.e., it should include important phenomena and should leave out the unimportant ones. This exhortation begs the question of which is which, but anybody who has done analyses in live contexts knows well the pressure from the client and the critics to include more and more detail in the model, and the importance of resisting many of these pressures. One of the difficulties in making models useful in practice is that a problem always contains aspects that can be blown out of proportion by word pictures and one-of-a-kind anecdotes. It is often necessary to resist model expansion in order to prevent the project from becoming too large and unwieldy and to avoid expending excessive resources on activities that will not really affect the results. There are often tough calls; to make good model design decisions you need side analyses and arguments and also organizational processes that bring the clients' inputs and judgments into play.

Two other important problem-solving paradigms that make heavy use of models are decision analysis and soft systems methodology. Decision analysis is a whole field dealing with problem structuring and decision making. See, for example, Keeney and Raiffa,[7] Howard[8] and, for a recent update on theory and practice, the special issue of *Interfaces*[9] on decision and risk analysis. Soft systems methodology[10] consciously integrates problem solving into organizational processes. Without describing these methodologies in detail we note that they too set contexts for model building in which relevance, understandability and sufficiency for the task at hand are more important criteria for judging a model than the classic scientific goals of generality and accuracy.

WHERE DO MODELS COME FROM?

Some people seem to sit down and produce an insightful and relevant model with apparent ease. How do they do it? Much has been written about problem solving. Smith[11] reviews and discusses a selection of this work.

If I were to give a personal, one word answer as to where my models come from, I would say: dissatisfaction. When working on something I often feel discontent at the state of my understanding — a conviction that something is not right and that I understand part of what is going on but not enough. I need to think about the problem more, work out further details and express them, verbally and mathematically.

This description fits quite well with certain standard ideas about problem solving — that we encounter a difference between our goal and our present state and try to reduce it. The view of differences as a driving force has a long history. Newell, Shaw, and Simon[12] use it in their General Problem Solver. It is basic to Pounds[13] theory of problem finding. And, of course, it has roots in feedback control, which goes back at least to Watt and his governor for a steam engine.

These ideas can be embedded in a process. When I am building a model, I often find it useful to go through a series of fairly well-defined steps. The first is to write down what I want the output to look like. If the model is supposed to produce computer output, I may sketch a list of output variables or even a screen display in the form that I would like the results to take.

As a second step, I look at the outputs and write down the inputs I think will be required: what data, what relationships, what values for parameters, etc. Just writing down the inputs, of course, implies that I have some sort of rough notions about key phenomena and cause and effect. Where do these come from? From past experience or general knowledge about the subject, often gathered by interviewing people who know about the problem and by reading background material.

Then, having written down outputs and inputs, I have created a difference between them that needs to be closed. Thus, the final step is to work on the model — what assumptions are required to take inputs into outputs, what functional relationships are needed. At this point we are on the way. The process goes through iterations. Ideas change about what the outputs should be. Inputs are never right the first time, and the steps, assumptions, and relationships in between are subject to much learning and development. However, the basic procedure works well in many situations. The reason is that the outputs conjure up in one's mind ideas about what inputs are required and inputs and outputs together push one to hypothesize relations between them, and so the development of the model moves forward rather naturally.

One thing that the psychological literature makes clear is the importance of prior structures. Thus, the concept of a linear model is imbedded in many of our heads so deeply that we are unaware of it. An archaeologist recently asked me to comment on a paper about diets of early American Indians. Without recognizing it as such, the authors were building a linear model of diet, but they lacked any background with linear equation systems. The result was an awkward and inefficient treatment of something that could be described and understood quickly by someone familiar with linear models.

Another prior structure for model building is decision analysis. Still another would be econometric models. The thinking and priors about solving a problem are likely to be quite different for a person trained in econometrics from another in decision analysis. Other priors might come from knowledge of mathematical programming, stochastic processes or system dynamics. When you have some of these structures in your head, you see them everywhere and bring along the intuition that goes with them, thus prior bodies of knowledge are places that models come from.

At the same time, however, prior models form a prison of preconceived ideas. This is often a subject for concern — one hears the complaint that somebody "has linear program — will travel". How can we break out of preconceived ideas? This issue has spurred much discussion. For example, in his book reviewing research on judgment and decision making, Bazerman[14] talks about heuristic-breaking heuristics and catalogs a few, brainstorming being a prototypical example. The idea is to liberate and enlarge your thinking space.

How I had my great idea. Some models turn out to be deeper, richer and more relevant than others and to break new ground. They are more creative. Creativity is another topic with much anecdotal data and considerable research. The mathematician Poincaré in a famous essay on mathematical creation[15] tells of climbing the steps of a bus and having a flash of revelation in which he saw exactly how to solve a problem he had been working on unsuccessfully for weeks. One might conjecture that aspiring mathematicians would start taking public transit, but actually, the 'aha' or 'eureka' phenomenon is well known and well documented. What happens is that people make an intense effort to solve a problem and the effort, even if unsuccessful, forms an important preparatory stage. This is followed by a relaxation and release of mental activity below the conscious level. Subsequently, key ideas often surface. (This was Poincaré's own explanation of what happened.) The formal description in Bazerman[16] identifies the steps as preparation, incubation, illumination and verification. Research has shown that at the verification stage many seemingly creative people fail to be truly creative, because the euphoria of the 'aha' is not followed by the hard reality of testing. I think many of us have experienced the euphoria stage only to tear our ideas to shreds a few hours later.

To summarize: Models favor the prepared mind. In building models we bring to bear heuristics and structures we have used before. Various check-lists and paradigms can help organize our thoughts. However, prior knowledge also tends to blind us to fruitful new directions in problem solving. Breakthroughs are facilitated by conscious efforts to bring new ideas to the problem, often in a sequence of intensive effort, followed by backing away, and, after a delay, the release of a fresh flow of ideas.

SOCIAL RESPONSIBILITY

Social responsibility and ethics are the topics of this workshop and although they will be dealt with at length by others, I have found myself thinking about them while preparing these general remarks. It is not clear that ethics in model building is fundamentally different from ethics anywhere else but the circumstances and manifestations can be different and dilemmas arise in new contexts.

We identify two stages and three actors in the process of problem solving with models. These combine in multiple ways to create fertile opportunities for misunderstanding and conflict of interest. The first stage is the interaction and knowledge transfer that takes place between the model builder and the immediate client. Thus, an OR/MS professional works for a manager, or a consulting firm does a study for EPA.

However, there is usually a second stage — between the client and a third party, the client's client. Let us call the client's client the supraclient. Thus, an internal OR/MS consultant may build a model for a marketing manager about whether or not to introduce a new product. Then the marketing manager will very likely turn around and take the results of the model to the president of the company, the supraclient, and ask for money to launch the product. In the public systems arena, a consulting firm may do a study for EPA. But then EPA may take the study to Congress. The second stage is an integral part of the value-creation process and introduces new dilemmas.

There are a variety of pitfalls for the model builder in these situations. Here are some:

(1) *The client already knows the answer.* What the client really wants is justification for an answer in hand, not a new answer. This is bad news for the modeler, especially when the client's desire was not conspicuous in the beginning. There is great potential here for ethical dilemmas. For example, the model builder, being employed by the client, feels pressured to produce the prescribed answer even if he or she finds it faulty.

(2) *The client presses the model builder for early answers.* Time pressure is no surprise; everybody wants results yesterday. However, sometimes preliminary results are delivered, and sometimes they are too preliminary. The client may prematurely take a position on the issue with his/her supraclient. Subsequently the model builder learns more and wants to change the results. This can be very difficult and create great stress.

(3) *The client does not really understand the basis for the model builder's results.* Consequently the client is uncomfortable about using them and is hesitant about taking them to his/her supraclient. This is a very common situation.

(4) *The black and white problem.* Both clients and supraclients would really like to have results with certainty — black and white answers. Let me paraphrase a characterization of managers and model builders from an earlier paper:[17]

Managers know only two probabilities, zero and one, whereas model-builders know all probabilities, except zero and one. Furthermore, managers act before they think, if they ever think, whereas model builders think before they act, if they ever act.

Another popular articulation of such managerial behavior is: "Ready, Fire, Aim".

One dilemma for model builders that arises out of these contrasting cognitive styles goes as follows: The model builder says: "Maybe I should just figure out what is right and then present that solution as black and white as I can." In other words, "Perhaps I should I do my analysis and come to my own decision about what is the best action and then simply emphasize that which will support the action and, without deliberately concealing anything, downplay evidence that doesn't favor the proposed action." The obvious difficulty with this approach is that the client's knowledge and values may not be accurately reflected in the final choice of action, so that the client's decision-making responsibilities have been usurped by the analyst.

(5) *Having the client provide judgmental input.* This is something I have advocated[18] and is, of course, everyday practice in decision analysis. However, if the client's judgments largely determine the results, people can ask: "Is this a sell-out that uses scientific trappings to push a preconceived solution?" Here we again need to recognize the two stages of the process. It seems reasonable for a client to put in judgmental inputs when the client is also the decision maker and a principal gainer or loser from the outcome — after all, it is his/her problem. Where things become sticky is when the content of the analysis is to be carried forward by the client to a supraclient. Suppose a model builder does an analysis of whether a product should be introduced and takes it to a marketing manager, but the latter does not make the decision alone. Now suppose the marketing manager puts extensive personal judgements into the model and then takes the results to the president to ask for resources to go forward. Does this create an ethical dilemma for the original model builder?

(6) *All models are incomplete.* The client probably does not realize this to the extent that the model builder does. The client may not understand too well what is missing and what is not and, in fact, really wants black and white answers in the first place. So what are the model builder's responsibilities?

Now it is my hunch that, if I were a playwright, I could take any of these dilemmas and make it into a good TV drama, showing some poor model builder caught between great forces, abandoned by friends, and ready to jump off the Golden Gate bridge with a UNIX workstation chained to a leg.

Ideas to Improve Social Outcomes

There are no panaceas, but here are a few suggestions that might alleviate some of the pressures. First of all, the front line is anticipation and prevention. Part of this is the management of expectations with clear prior specifications and criteria for evaluation of the model. A lot more is education, communication, and close involvement of the client — all vital yet expensive activities. The model builder should take responsibility for providing the client with valid, understood, problem solving information. As model builders we should understand what we have done and be confident about what we can or cannot say. This requires a whole lot of testing and evaluation as we go along and the communication of the essence of our results to the client.

On this score I have a two sets of questions for managers to ask any consultant, internal or external, who is building a model or doing an analysis for their organization. The first is:

(1) What is the answer? What are the conclusions of the study?

Given these, a mangager should ask: "Do I, as the client, believe these conclusions? Why or why not?" In other words (and many managers do this instinctively), clients should compare the results presented to them with their own mental models and intuition and should discuss the differences with the model builders. The second set of questions is:

(2) Why did the answer come out as it did? What were the key assumptions and data that determine the results? And what was the essential reasoning between input assumptions and output results?

Any good model builder should be able to answer these questions clearly in lay language and, indeed, should have asked them of him/herself. One of the tricky aspects of model building for problem solving this is that the greatest benefits lie in coming up with unexpected answers. Thus in favorable circumstances we shall have produced something different from the client's mental model. However, at the same time such answers require the most testing. But that's OK, it's the business we are in.

Related to the issue of conflicting goals of client and supraclient is the problem of multiple clients with quite different world views — for example, a marketing manager and a production manager for a decision involving a new product. Kusnic and Owen[19] describe an interesting approach to this problem: The analysts do two projects, one from each point of view, and have the participants understand both sets of results and negotiate the final decisions.

Another suggestion, which picks up certain themes in an earlier paper,[20] is that we should design models as much as possible to be transparent, that is, easy for the client to understand, with inputs and outputs expressed in everyday terms with readily interpretable units. Sensitivity analysis should be easy to do and the models should be robust in the sense that it should be hard to obtain non-sensical answers from them. We should also seek open and transparent validation, i.e., testing done in ways that are easy for the client to understand and ask questions about.

A further possibility is to introduce some form of peer review and open debate. This sometimes takes place automatically on public issues but in corporate settings is largely missing. Adversarial debate also has a downside, because, although devil's advocates are important, much good analysis is artful and has carefully left things out that should be left out and put things in that should be included. It is easy to conjure up phantoms and claim that effects not included are so important that their absence invalidates the results. Skepticism and critique are healthy, but we should not complicate and self-criticize ourselves into impotence when we have a lot to contribute. As stated earlier, our underlying responsibility is to provide the client with valid, understood, problem-solving information.

NOTES AND REFERENCES

1. Churchman, C.W., Ackoff, R.L. and Arnoff, E.L., *Introduction to Operations Research*, John Wiley, New York, 1957.
2. Urban, G.L., "Building Models for Decision Makers," *Interfaces*, 4(3), 1–11, May, 1974.
3. *Op. cit.*
4. *Op. cit.*
5. Kolb, D.A. and Frohman, A.L., "An Organization Development Approach to Consulting," *Sloan Management Review*, 12, 51–66, Fall, 1970.
6. Little, J.D.C., "Models and Managers: The Concept of a Decision Calculus," *Management Science*, 16(8), B-466-485, April, 1970.
7. Keeney, R.L. and Raiffa, H., *Decisions with Multiple Objectives: Preferences*

and Value Tradeoffs, John Wiley, New York, 1976.

8. Howard, R.A., "Decision Analysis, Practice and Promise," *Management Science*, **34**(6), 679–695, 1988.

9. Volume 21, No. 6, November–December 1992.

10. Checkland, P., *Systems Thinking, Systems Practice*, Wiley, London, 1981.

11. Smith, G.F., "Towards a Heuristic Theory of Problem Structuring," *Management Science*, **34**(12), 1489–1506, December, 1988.

12. Newell, A., Shaw, J.C. and Simon, H.A., "Report on a General Problem-Solving Program," *Proceedings of the ICIP*, Paris, June, 1960.

13. Pounds, W.F., "The Process of Problem Finding," *Industrial Management Review*, **11**, 1–19, 1969.

14. Bazerman, M.H., *Judgment in Managerial Decision Making*, John Wiley, New York, 1986.

15. Poincaré, H., "Mathematical Creation" in James R. Newman, *The World of Mathematics*, Vol. 4., 2041–2050, Simon and Schuster, New York, 1956.

16. *Ibid.*

17. Little, J.D.C., "BRANDAID: A Marketing-Mix Model, Part 2: Implementation," *Operations Research*, **23**, 656–673, July-August, 1975.

18. *Ibid.*

19. Kusnic, M.W. and Owen, D., "The Unifying Vision Process: Value beyond Traditional Decision Analysis in Multiple Decision Maker Environments," *Interfaces*, **22**(6), 150–166, Nov–Dec, 1992.

20. *Ibid.*

Chapter 4

Part 3. Morality and Models

Richard O. Mason

A GROWING NEED

Model building, once the purview of an elite set of management scientists, is now the province of many. The personal computer and programs in business and engineering schools serve to democratize its use. During the last decade or so, many college graduates have been exposed to the rudiments of model building, and most of them have access to personal computers capable of performing the necessary calculations. Most also have software that greatly facilitates their work. There are, for example, over 5 million copies of Lotus 1-2-3, the most popular spreadsheet language, in use today. All of this adds up to a fairly widespread diffusion of model-building capacity.

Just two decades ago, a modeler's choices generally boiled down to turning simplex pivots by hand or standing in line to submit a Fortran program deck to a mainframe computer. Today the model builder is likely to have a machine with a 286, 386, or 486 chip in it that places more power on one desk than a whole room contained before. If the model builder works in a corporation or government agency, a management scientist's personal computer or workstation is likely to be connected to a large integrated database that provides data about widespread aspects of the domain being modeled. And the model builder may be able to share information with others by means of a local area network. Consequently, data collection, once the nemesis of the practicing management scientist, is often now reduced to down loading from a mainframe or exchanging a disk with another person.

The techniques of model building have indeed been democratized, but one must ask whether the ethics of model building have followed suit. Do today's many and varied model builders assume professional responsibility for their models? The question is crucial, since a special fiduciary relationship exists between the model builder and his or her users or clients.

At a minimum, a model builder is obliged to do at least three things: (1) to represent reality to clients adequately, (2) to understand and to incorporate the clients' values into the model in an effective way, and (3) to insure that actions the client takes based on the model have the desired effect. These are the three major covenants a management scientist or operations researcher has with clients, and they comprise the core of the ethics of the profession. Unfortunately, these covenants do not come sealed in a cellophane wrapper along with every software disk. Rather, they must be learned, as all moral principles must, through reflection and by means of the assumption of responsibility. Most senior practicing management scientists have learned these lessons from experience and have made them a part of their professional bearing. New modelers must now learn these lessons.

A CASE FROM THE PAST

In the early 1970s, West Churchman and I began a brief project with the Boise Cascade Company. We had received a grant from NASA to explore various ways in which information obtained by the soon-to-be-launched Earth Resources Technological Satellite (ERTS) might be used by managers in government and industry in their planning processes. We learned about an operations research (OR) group at Boise that was in the process of putting together a large-scale, multistage, decomposition linear programming model designed to aid long-range decision making for the company's timber management and paper production. The model represented a 100-year flow from forests to final product demand, including cutting trees, planting, reharvesting, felling, transporting, peeling and barking, sawing, and distributing products such as lumber, plywood, pulp, paper, and paper products. West and I believed that the forest inventory data available from ERTS would be useful to the planners and that by running the model we might be able to estimate the value of this more accurate information for the Boise Cascade Company and for similar forest and wood products companies. This was our initial plan, and to some extent, it was also the OR team's plan. Subsequent events, however, overtook by far our ability to carry it out.

The first indication of an impending ruckus came during a phone call from one of the timber personnel. He wanted to tell me that his people had decided to cooperate with us after all. Now this was surprise to me because I never knew that they had refused to cooperate in the first place. I did know that Boise Cascade was organized into two mammoth divisions: (1) lumber and timber and (2) paper. I also knew that the paper division had been long-time users of operations research, having had significant successes with projects such as optimal paper cutting, plywood composition, and allocation of production to plants. We were told that the management scientists at the paper division were very supportive of the corporate effort to develop an integrated long-range plan because, as one executive put it, "They are believers in models".

We were also aware that most of the people in the lumber and timber division were not "believers". They had few, if any, OR successes under their belt. More importantly, they came to their part of the business from a different philosophical point of view. They were trained as foresters, and most believed deeply in ecology and its values. Their style was to plod through forests in round-toed boots and plaid jackets marking trees and taking extensive notes in little spiral-bound books. Analytical models, to their way of thinking, were rather beside the point. Thus, we did not expect a groundswell of enthusiasm for our project from the lumber and timber people; but we had no reason to believe that they were the "enemy" and would fight "tooth and nail" against the corporate effort. My informant, however, told me that was exactly what had happened and that executives from the lumber and timber division, wary of a corporate attempt to usurp their control over timber decisions, had lodged a complaint against the model-building effort and had openly refused to support it. Somehow in the meantime they had changed their minds and now, he told me, "Everything was O.K".

"What's going on?" I wondered. During our next visit to Boise, the plot began to unfold. It had a long history that began with the company's founding. The core of the modern corporation stemmed from the old Valsetz Lumber Company. Timber and lumber products were the heart and *raison d'etre* of the original business. The paper business was a latecomer. A binge of mergers and acquisitions and a major reorganization had ultimately resulted in the new division corporation with headquarters in Boise. In the new organization, both divisions, but especially the paper division, were placed on heavy profit incentive plans, complete with cash and stock bonuses. Thus, higher divisional profits meant more money in a division executive's pocket. Executives in the paper division were reaping the harvest of this incentive program, a fact we soon learned was inadvertently publicized by the planners while they were constructing the model. It became the basis for a major corporate civil war.

The battle was joined at the estimate of the cost coefficient to be used for transferring bark and chips from the lumber and timber division to the paper division. The chips are chucked out when logs are debarked and peeled prior to being sawed into lumber. Before the paper business was established, these superfluous chips were burned in large kilns found throughout the states of Oregon, Washington, and Idaho. Chips in those days actually had negative value, since it cost money to burn them. Hence their effective transfer price was negative or at best "zero". The paper business came as an economic godsend, since it took this virtually worthless by-product and turned it into something of value.

In time, the economic value of the paper business became greater than that of the lumber business. Throughout the years, however, the effective transfer price of chips remained low, hovering near zero. When the corporate modelers came upon the scene, the accounting price that the paper division paid the lumber and timber division was, say, US $1 a unit.[1] The long-range planning application of the model, however, called for the examination of a wide set of alternatives, and as the model builders explored the possibilities, they found that the economic pricing of chips was very illusive. Indeed, they were faced with a quandary. It went something like this:

History placed the transfer price at zero. Present accounting practice placed it at US $1. Yet if the entire log was peeled into chips and its full cost embodied into the cost of chips consumed, then the price would be closer to US $8. To further complicate matters, the Japanese were willing to pay as much as US $10 for chips in the spot market. So what was the "real" price? By asking the question, the corporate model builders had stumbled into the middle of an unspoken but deeply held social pact. For years a tacit agreement existed between the timber and paper people as to the cost of chips, a fundamental resource in their businesses. It had become a sort of cultural norm. Once the norm was exposed and put on the table, the timber people began to feel that the practice had been unjust. So smitten, they had wrung their war bells and begun to fight in earnest.

As the battle ensued, the once anti-OR lumber and timber people became converts, almost overnight; and for good reason. They stood to gain as much as an 8–10 times improvement in their sales price, providing a big boost to their bonuses and more than tiding them over the slump in lumber demand they were experiencing at the time. Meanwhile, the formerly pro-OR paper people quickly became antagonists to these models because they could see immediate losses in their divisional, and hence personal, pocketbooks. As the combat escalated, Robert Hunsacker, Boise's chief executive officer at the time, stepped into the middle of the melee and called a moratorium on model building. He reaffirmed the existing US $1 accounting transfer price and ordered a discontinuance of the corporate model-building effort. The planning model was then essentially scuttled.

Within a few weeks, several of the key corporate OR personnel had found jobs elsewhere. In an interview, Hunsacker told me that "They (the corporate OR team) may have been right, but if we had gone down that track, it would have destroyed the company". He was likely correct in his assessment.

SOME LESSONS LEARNED

What does this story tell us about the ethics of models? It describes what I believe is the greatest quandary any model builder faces. On the one hand, every management scientist has a fiduciary responsibility to represent reality in his or her models and to do so with as much fidelity as is possible. The special knowledge of model building that is the *raison d'etre* for the profession includes examining a problem situation, framing it in terms of a model, collecting data according to the model's variables and parameters, and deriving conclusions. The model builder then asserts that within the limits of the method and its assumptions the model adequately represents reality. The model builder has a mandate to produce a valid representation. This is part of the trust relationship he or she must establish with those who use his or her models.

Similar to a physician, however, a management scientist must also agree to serve the best interests of the client. The other form of arcane knowledge a model builder possesses is the ability to determine the client's objective function and to incorporate the client's values into models in such a way that *when the client acts upon the conclusions of the model the results are as predicted and are beneficial to the client.* This is the second part of the covenant made between a model builder and the client.

So the management scientist qua professional enters into two major implied covenants that define his or her primary fiduciary responsibilities. These are: (1) the covenant of reality, in which the model builder is entrusted with understanding things as they actually exist in the problem area and representing their most salient features as accurately as possible, and (2) the covenant of values, in which the model builder is entrusted with the visions, goals, and objectives of the client and pledges to serve those values as loyally as possible. For the most part, management scientists have little difficulty adhering simultaneously to these two covenants. Sometimes, however, they come into conflict. When they do, a management scientist is faced with a moral dilemma.

The corporate OR team at Boise Cascade were avidly pursuing their covenant with reality. Their compelling question was, "What is the 'true' cost of the chips?" However, they were initially out of touch with their covenant with values. What were the values of cohesion for the organization? What were the overarching goals of the executives? Who was being served? The team members posed the true cost question and pushed for an answer. They never considered the second set of questions until, in fact, the answers sought them out and made their presence known. That is, the corporate value of keeping the organization together may well have exceeded that of determining the most accurate cost of chips. The OR team only discovered this when Hunsacker called his moratorium.

Whenever a management scientist takes on a new project, he or she should, just as a lawyer should, determine two things. Can I protect the rights of my client(s)? And, can I pursue the truth? The answers to both of these questions can lead to awesome responsibilities.

A COVENANT WITH REALITY

The model builder asserts that the model he or she has designed adequately represents reality. In order to do this, the model builder first selects a modeling language, one based on a theory such as economic order quantity or linear programming theory. Then, within the structure of that language, the model builder crafts a specific model. Several important decisions are made at each step of the process, as a historical example will serve to illustrate.

One of the first theoretical structures used for general-purpose model building was an application of probability theory made by A. K. Erlang in about 1910 at the Copenhagen Telephone Company. His creation came to be known as queuing theory. People making telephone calls, he assumed, do not lift up the receiver at a precisely scheduled or predetermined time. Rather they place calls at "random". By creating a probability distribution of incoming calls and also by determining the characteristics of the operator and the equipment — known as the "server" in queuing theory — Erlang was able to deduce such properties of a system as the average time customers waited before their call was placed, the average number of customers in the waiting line, and the average idle time of an operator. Recalculating the model for various numbers of operators revealed the response of these output variables to the number of server stations.

A typical simple queuing model contains just three parameters: (1) the total time of operation (expressed, say, in days or hours), (2) the mean arrival rate of inputs, and (3) the mean service rate — the time it takes for a station to perform its work on an input and issue it into the output stream. This rather simple model, however, yields a rather substantial set of inferences about the system's average or mean performance, including the number of output units in the queue during the time period, the time each waits until it is serviced, the time that the service station is idle, the number of idle (or busy) periods, the time an input stays in the system, and the number of inputs served per busy period.

What is <u>not</u> included in Erlang's model, however, is as important as what is. The content of the incoming calls, for example, is assumed to be irrelevant. Two lovers making contact after a long absence are given the same weight as two nosy neighbors exchanging the latest gossip or two executives trying to do a deal. Teenagers catching up on the day's events are given the same weight as people in distress attempting to place an emergency call. Also ignored are the frustrations of unserved customers waiting to make a call and the irritations of harried operators trying to cope with a flood of incoming calls. In short, Erlang's model treats the people in the system as if they were machines. It admits no humanistic factors.

Among the moral questions raised by applying this model are whether these three parameters adequately represent the reality of the situation for the client and, of course, whether the client is aware of and understands the exclusions. This makes the model builder's intellectual task quite difficult. From among a myriad of possibilities, he or she must choose just a few factors to represent the nodes of the system. From among the vast numbers of possible relationships that might exist between these factors, he or she must choose just those that he or she believes best typify the relevant structure of the system. The ethics of model building require that the model builder make every effort possible to match the elements of the model with corresponding elements of the reality and to appraise the client as to where the significant differences occur. This is the essence of the model builder's covenant with reality.

Matching a model with reality becomes an exceedingly difficult challenge when the problems are fraught with complexity, as they almost always are. Most situations are "messes", to use Russ Ackoff's apt description. Models, on the other hand, are by definition rational and analytical. So the management scientist generally must try to tame the "wickedness" of the mess so that it will fit adequately into the model. All too frequently, however, the management scientist sides with the mathematical aspects of the model because this is perceived to be "rigorous", "objective", and "scientific". Thus the dishevelment of the mess is ignored. Kenneth Boulding writing in *The Image*, described the process well: "With deft analytical fingers the economist (read model builder) abstracts from the untidy complexities of social life a neat world of commodities".[2] By neglecting underlying complexities, a management scientist reneges on the covenant with reality and places the entire profession in jeopardy.

As the world has become more complex, model building has become a more secular activity. A vast number of people now engage in model building of some sort. As a consequence, adhering to the model builder's covenant with reality has become very difficult. For many of these neophytes, the model *is* reality, not a carefully crafted prism through which to view it. All modeling, of course, entails the construction of a simplified intellectual form that permits explanations and inferences to be drawn and measurements to be made. The difference lies in how the form is used. The skillful management scientist turns the facets of the model, striving with each permutation to glimpse more deeply into the underlying reality. At each turn, the model builder conducts a dialogue between the reality of the problem and its portrayal in the model, noting differences and similarities and assimilating insights. After each of these exchanges, the model is refined, either in its concrete specifications or in the model builder's subtle understandings of its meaning. As the dialogue progresses, the model moves toward reality. It never reaches it, of course — that would obviate the need for a model — but with each successive movement in the dialogue the management scientist becomes more "at one" with the model and with the reality of the problem he or she seeks to solve. At some point, the model builder comes to the professional judgment that the present form of the model adequately represents the reality. He or she knows what the model incorporates and what it does not. For example, if a lovers' quarrel is important, it may not be explicitly represented in a telephone system queuing model, but it will be contained in the intellectual tapestry the model builder weaves around the model. This expanded and deepened understanding can be communicated to the client, enriching the client's understanding of the problem as well. When this total communication process is complete, the model builder has complied with the covenant and with reality.

A CASE OF A COVENANT WITH REALITY

In the late 1960s and well into the 1970s, Harry Grossman directed a very successful operations research effort at Security Pacific Bank in Los Angeles. The range of project areas examined ran from check processing to lending practice. Virtually no part of the bank was left unexamined, and most yielded to improvements when his team of management scientists went to work on them. With each success, the team moved a little closer to the corporate board room. Finally, as the bank grew along with California's thriving economy to become one of the largest in the world, Harry was asked to lay out a master plan for locating the bank's new branches and, in particular, to determine the best location for its new world headquarters. This master plan was to become an integral part of the bank's strategy for decades to come.

The team began to collect reams of data about business development, real estate values, population trends, and all of the manifold activities that influence location decisions. They interviewed executives inside and outside the bank extensively. A prototype model was constructed and a computer run completed. The team pored over the model's results, arguing, debating, clarifying. More data was collected and more interviews were conducted, guided, of course, by insights gained during their deliberations over the prototype model. The model was refined and run again. More discussion. The process was repeated again.

After several months, the team concluded that it had completed its job. A final run of the model was made and the results presented to the bank's senior management. The model showed clearly that Century City, near the old Twentieth Century Fox studios, was the optimal location for the new headquarters. The executives were impressed and engaged Harry in extensive discussions about possible locations for various new branches and operations facilities as well as focusing on the central issue: where to locate the headquarters. Everyone seemed to agree with Harry's arguments. By all indications, the model had been a sterling success.

Thus, Harry was stunned a few weeks later to learn that the bank had decided to place the new headquarters on Bunker Hill as part of the redevelopment of old downtown Los Angeles rather than to build at Century City. He was puzzled and felt more than a little let down. His disappointment was soon assuaged — somewhat — when Carl Hartnack, then CEO and Chairman, called him into his office to congratulate him on the model and to tell him how valuable he believed the management science group's efforts were to the bank. Hartnack went on to explain that the Century City location was undoubtedly the best location, given traditional banking and economic factors, but that the decision had been made on considerations that ranged beyond these. The factor that finally tilted the decision in favor of the Bunker Hill location was that the bank felt the need to participate in the redevelopment of the central city and to play its part in removing some of the urban blight that was rapidly encroaching upon the downtown area. Harry's model, he said, provided him and the board with a clear sense of the additional cost the bank would incur by doing what it believed to be its civic duty. Indeed, the results of the model had been very instrumental in helping them make their final decision.

Several years later, as he was preparing to retire, Carl Hartnack told Alan

Rowe and myself that the management science location model was probably the most valuable management information project ever carried out at the bank. For nearly a decade, he said, no location decision had been made without playing it against Harry's model and usually discussing it with Harry himself. It "captured the central elements of banking", he said.

To be responsible to reality is not the same as copying it. Harry's location model was *not* reality. It failed to take into account such things as urban renewal. However, it converged toward a relevant reality in a way that bankers and management scientists could understand and it revealed important insights to them as they immersed themselves in it. These bankers, all non-management scientists, could readily add the missing ingredients to the model and easily comprehend their implications. In the end, Harry Grossman's dialogue with his model was truly successful. By struggling to incorporate as much of the relevant reality as possible at each step along the way, he produced a template that captured the mind of his clients in a forceful and fitting way. As a professional, he kept his covenant with reality.

A COVENANT WITH VALUES

Erlang's queuing model did not specify how long a customer should have to wait to get a free telephone line or how long operators should be idle waiting for calls to come in. These were managerial objectives. The values that underlie them were not made explicit in his model. Rather, he assumed that the user would apply the appropriate values to the model's results after the fact. This was common practice in model building during the early days of the profession. Development in mathematics, however, changed this. The two major innovations were the economic order quantity (EOQ) model published in 1915 and T.C. Koopman's and George Dantzig's development of the linear programming (LP) model during the early 1950s.

The calculus of the EOQ model and the matrix algebra of the LP algorithm find solutions that minimize costs, maximize revenues or, in general, "optimize" some objective function. With these new techniques the question of values and how to identify, measure and incorporate them was brought more forcefully still into the building process. With the advent of these new model-building theories, management scientists assumed a new obligation: to capture their clients' values accurately and to reflect them in their model's objective functions.

In the typical EOQ model, as Churchman has pointed out, one must introduce data such as the demand for inventory, the cost of ordering new stock, and the cost of holding inventory in storage. This latter factor, inventory holding cost, is especially laden with values. The sacrifice a businessperson makes because he or she used money to buy inventory rather than do other things depends totally on the person's values and on the options available. The person could have, for example, put money into a real estate limited partnership promising a 77% return on investment or into a certificate of deposit guaranteeing 8%. The person certainly might have reinvested it in other parts of the business. Or, feeling lucky, the person might have placed the whole wad on "Galloping Light" at Churchill Downs. Whatever the client's values, the management scientist must understand them before he or she can determine an inventory holding cost. It is a moral obligation as a management scientist to do so.

The process of values discovery can lead to daunting moral dilemmas. What does a management scientist do if his or her conscience will not abide the client's values? There are two schools of thought. One is that a management scientist is *a scientist* first. Therefore, he or she owes allegiance only to the tools and techniques of the trade. The model builder's responsibility, so this position goes, is to insure that the model is built and executed with scientific precision using the best methods available. He or she is not responsible for what the client's values are or what client does with the model. Members of the other school take strong exception to this. They argue that a management scientist is a *human being* first and a scientist second. As a human being, the model builder must look beyond the process of model building itself and take responsibility for the total impact on human kind. Management scientists who believe in this second position have refused to work on projects with goals contrary to their values. Examples include refusing to build models for military systems because the analyst's values abhorred war and killing and refusing to help build a national databank of citizen information because the analyst was concerned about the temptation by overzealous politicians to make abusive use of the information. Also in this latter school was Loren Cole's consulting associates in Berkeley. The group would only take on projects that were consistent with their own personal values, such as their desire to help solve the world hunger problem. Their first step with a client was to conduct a value audit. If there was no agreement, the team departed. If the client's and team's values coincided, however, the team attacked the project with great zeal and enthusiasm, often beginning by writing proposals to third parties soliciting funds for the project.

Keeping one's covenant with values is often more taxing than keeping one's covenant with reality. It involves deep introspection, the coming to grips with one's own values as well as those of others. It also requires the model builder to translate values — which are generally abstract — into the concrete terms of the model. Most of us are not well prepared to do this.

CONCLUSIONS

Model building is the core process of the profession of management science. It is a tool of enormous power that should be directed and controlled. Management scientists wield this power and are thereby responsible for its use and outcomes. As professionals, management scientists make two important promises — covenants — with their clients and with society at large. They promise to represent the real conditions of the problems they are solving with as much fidelity as possible, and they promise to secure improvement in the client's world by adhering to the client's values with as much fidelity as possible. Since a management scientist possesses special knowledge and skills for building and interpreting models and, frequently, exceptional technology in the form of computers, software, and data sources to execute models, the keeping of these promises creates a relationship of trust. Since the client does not possess the knowledge and technology of model building, the client must trust the management scientist to keep his or her promises. This is the essence of the fiduciary relationship between the two.

The trend toward democratization of information technology and model building, characterized by the proliferation of faster personal computers with more memory and ever more powerful software packages to go with them brings with it a new challenge. How will these new model builders who increasingly possess the knowledge and technology of the profession be inculcated with the values of the profession? Will the covenants, so central to the practice of a seasoned professional, be kept by those who are just entering the field? These are the new questions about morality and models that are facing us as we prepare for the third millennium.

NOTES AND REFERENCES

1. My memory fails me now as to exactly what the transfer price was. If I could recall the exact figure, it would likely still be confidential. The relative values I have quoted, however, are as I recall them.
2. Boulding, K., *The Image*, Ann Arbor Paperback, The University of Michigan, 82, 1961.

Chapter 4

Part 4. One Sided Practice — Can We Do Better?

Jonathan Rosenhead

INTRODUCTION

Those who model decision situations have worked almost exclusively for one type of client: the managements of large, hierarchically structured work organizations in which employees are constrained to pursue interests external to their own. Other organizations abound in any developed society, disposing of few resources and operating by consensus rather than chain-of-command to represent interests that unite their members. The problems of this alternative clientele are no less taxing than those of larger organizations. Complexity and uncertainty characterize their environments, not least because of their vulnerability to the operations of the modelers' conventional clients. Recent developments in problem-structuring methods are providing a repertoire appropriate to such situations. Can our profession respect itself, and so expect the respect of others, if we aid the powerful and neglect the weak?

A caveat to start with. No two individuals' experience and assumptions are quite the same. The transferability of British experience across the Atlantic is therefore clearly problematic. Its validity may be restricted to its context of origin, or it may take on a different significance elsewhere. Certainly the British modeling practitioner operates in a corporate environment different from that in the United States, and the British university is an institution quite distinct from its American counterpart. It is unlikely, then, that "lessons" from Britain can be advanced with confidence for straightforward application elsewhere. It is not just that there are no universal answers; it may turn out that we are asking different questions. There is certainly a risk of engaging in non-intersecting discourses or parallel rhetoric.

That is the end of the disclaimer. The rest of this chapter will make no apologies.

THE CONSEQUENCE OF DIFFERENTIAL MODELING

It seems to me to be necessary to take a view of ethics broader than the moral choices of the individual practitioner. Such choices, if made after principled introspective deliberation, can undoubtedly make the individual feel better. However, the practical effect for anyone else may be minimal, unless such admirable behavior becomes quite general, which it won't in a society that inculcates individualism, encourages selfishness, and rewards greed. "If I don't do it, someone else will" is only too true. Unless the concept of ethical behavior is for individual gratification only, it must be based on the *consequences* of action.

Social responsibility is not attainable in the head or even at the hands of the individual practitioner. John Mulvey[1] suggests that significant change in this respect is only likely through social pressures from stronger forces outside the profession itself. I would not go quite so far. I believe that modelers can be part of the solution, whereas at present we are part of the problem. However, that solution cannot be purely voluntaristic and individualistic.

The scale of the problem and its systemic nature require an organized, cooperative component to any effective strategy.

If our conception of ethics is to be based on the consequences of action, the notion of "harm" is clearly central. Actions that may cause harm to others, whether deliberately, carelessly, or recklessly, are those that must be subject to question. No harm, no problem. But the notion of "harm" raises that of "power". A possibility of harm exists only where one individual (or group) has a measure of power over another. Power over another is the ability to achieve outcomes for an individual that he or she would not have chosen for himself or herself.

How does the modeler exercise power? It is not possible to go into this issue here at length, but it is not necessary either. If one does not believe that models are efficacious (that is, useful in principle and often in practice for their users) then the issue of ethics and modeling is of no practical significance and so of no interest. If the models are useless, then they can harm no one but their users, and the current class of users has the clout to purchase good advice or to absorb the losses if they fail to take prudent precautions.

In a modeler, such a view would be downright cynical, but it is also unconvincing. Without models, decision making is too often left to intuition and hunch, which in a complex and changing society can encounter insurmountable roadblocks. Debate descends to assertion, and both process and product suffer. That is why the use of models is a source of power. The "user," with an appropriate model, is able to scheme more effectively and to this extent is able to secure more favorable outcomes. (In a conflictual society, this in many cases means outcomes less desired by some others.) Since the model could not in general[2] be developed or applied without the assistance of a modeler, he or she is a necessary element in the empowerment of the client.

However, it is not the power advantage of one organization over another of comparable resources which, if it had the wit and the will, could have commissioned comparable research that need concern us. Such moral hypersensitivity would be quite practically disarming. It is, rather, those cases in which negative consequences occur for individuals or groups with little power to resist, to which ethical considerations apply.

Models are not generally paid for by one party for its purposes and then made voluntarily available to those who might have an interest in obstructing the resulting developments. In any case, models are not equally useful to all parties. They are normally designed to capture those elements of a situation that are relevant from one world view for a client with particular resources at his or her disposal. For those who do not control those resources and whose world view incorporates factors not present in the model, the model might have little or no value even if it were made available.

This aspect of modeling work was captured by Peter Szanton, who led the New York City Rand Institute, when he wrote, "Work such as the Institute's affects the balance of power as between government and governed...... I cannot say that our work has been much affected by this fact".[3] We will return to this quotation, and supply the two omitted central sentences, a little later in the argument.

Empowering the already powerful, and thereby allowing them to harm the interests of others, is one aspect of modeling's ethical situation. However, "harm" is a relative concept. Another way of harming people is by refraining from doing good. If we inhabit an inequitable but potentially improvable society and simply fail to participate in its improvement, we thereby visit harm on the disadvantaged relative to the improved conditions that might have been attainable with our help. In particular, if models that might have been constructed, which could have been of advantage to the disadvantaged, are not so constructed because of the absence of a modeler, an ethical question is clearly raised, especially if the modeler is doing well because of the status quo. Of course, the modeler may not notice or be aware of the disadvantaged or their disadvantage. However, in societies such as ours, this must constitute ethical carelessness on a grand scale.

MODELING, MOTIVES AND MANAGERS

Certainly the (British) founders of operational research (OR) did not adopt so narrow an approach. They are celebrated, if at all, in the opening pages of all those college texts as brilliant scientists, bringing their habits of observation, hypothesis formation, and validation to combat irrationality and inefficiency. What is not usually reported is that they were motivated to a significant extent by the wish to make the world a better place.

To be more precise, many of the leading members of that pioneering wartime OR community, and a fair sprinkling of the less famous, were socialists.[4] They had been participants in the 1930s movement urging the socially responsible use of science. The key intellectual figure in this movement was J. D. Bernal, himself an early operational researcher, who held that science was a progressive force whose full potential was prevented from being released under capitalism by the restriction of the profit motive. By pressing for science to be used in the public rather than sectoral interests, Bernal held, scientists would be helping to show up the limitations of capitalism and bringing the progression to socialism nearer. It was, disproportionately, people motivated by ideas like these who took the wartime opportunity to do science in the public interest.

Of course, the predictions that capitalism would be unable to make whole-hearted use of science have been refuted by postwar developments. Indeed, deliberate, planned technological innovation has become one of the elements in the economy's internal dynamic. The ideas of the early socialist operational researchers were decisively swept away by the laws of gravity of contemporary society. Ideas need sponsors capable of adopting them and putting them to use. The only takers then in sight for the idea of a self-conscious activity of rational analysis of organizational problems were the management of major organizations.

These dangerously radical elements of our heritage as modelers have, of course, long been expunged from history. For much of the nearly half century since then, it has been hard to discern such influences in current practice. Indeed, the modeling profession, with OR at its center, has operated with an ever-narrowing focus. Our activities have displayed, in fact, an ever-more-frantic quest for managerial acceptance. This is not, of course, a wise policy even in its own terms. Hanging on desperately to the coat-tails of the powerful is both an uncomfortable and an undignified posture.

The result of this policy is not the unqualified managerial accolade that its exponents might have hoped for. However, it has nevertheless produced a bias in the practice of modeling so complete and obvious that most of the time its existence is not even noticed by modeling practitioners. The bias is that, although advanced societies are populated by myriad variegated organized groupings, we work for only one type.[5] These are what Ackoff[6] called "uninodal homogeneous organizations". A uninodal organization is one with "a pyramid of authority topped by an ultimate decision maker". A homogeneous organization is one "that has greater control over its members than its members have over it". In more common parlance, these are organizations in which people work for a wage, under the direction of a management hierarchy, in pursuit of objectives defined at the top.

To take one example, modeling work is rarely carried out for trade unions. Consider: as a member of the Program Committee of the 1976 Conference of the (British) Operational Research Society, I found myself taken to one side by the Committee Chair who spoke gravely to me about the need for "balance" in the program. I realized that he was concerned about the lineup for the stream of papers on "Participation" that I was organizing. The nub of the matter was that two of the speakers were trade unionists. Of course, there were seven other speakers in my stream. There were no trade unionists at all in any of the other four streams. Indeed, I doubt that one could have found more than two trade union speakers in all the 18 previous conferences put together. Yet it was my inclusion of trade unionists, not their blanket exclusion elsewhere, that raised the question of "balance".

The bias in modelers' clientele was borne on me again more recently when I addressed the management science group of a major firm of accountants operating in both Britain and the United States. To help me prepare my talk, they sent me a copy of a promotional brochure that listed, *inter alia*, some of their current or recent clients. I discovered that these happened to include my bank, my mortgage company, the supermarket where I shop in some discomfort on alternate Saturdays, the railway running the train that got me to the meeting late, and the telephone company whose bills reach a magnitude that my bank doesn't like. These are all organizations with which I have problematic relations. Though I don't work for them, I certainly don't experience *them* as working for me. Indeed, I experience them as opposing forces.

On the other hand, there were a whole string of omissions from that client list. The firm was *not* working for the National Council for Civil Liberties, for my trade union, for my school parent–teacher association, for the Labour Party or the Green Party, for the Friends of the Earth, or even for my local Borough Council. Most of these I feel more positive about than the groups that had commissioned the firm's advice.

Of course, a response might be that the firm would not work for most of these organizations because they couldn't afford to pay. That would not be accepted, however (at any rate in Britain), as a valid argument for excluding whole classes of clients from access, say, to medical or legal services. Doesn't the modeling profession think its services are really important? Another response is that such organizations are political and so to be eschewed. Yet this firm was at the time heavily involved in work connected with the British government's program of privatization of publicly owned utilities and other assets. This program was under severe critical assault on technical grounds, politically controversial, and deeply unpopular. Nevertheless, the firm did not find this environment too "political" to stomach.

AN ALTERNATIVE CLIENTELE

In a world of power inequalities and conflict, in which self-interest may seem at odds with altruistic motives, it is hard to find ethically firm ground. Churchman,[7] writing about operational research as a profession, quoted a principle from Kant as one possible basis for ethical action: "Make only those decisions which treat humanity (in you or in another) as an end, never as a means only". This does create a paradox (to put it no higher) for practitioners of OR and related disciplines. For it is difficult, Churchman observes, to conceive of an OR recommendation that does not treat some people as means only. For example, cost reductions mean loss of employment. But more generally, OR collaborates with management in instituting changes in organizations without any necessary participation whatsoever on the part of the rest of the people concerned. There would surely be a wide measure of agreement today that what Kant calls "humanity in a person" includes that person having a strong say in the decisions that affect his or her life.

Kant's principle can be read in a strong or a weak version. The weak interpretation is that decisions are acceptable so long as they do not treat humanity as a means *only*. This leaves us in the clear, provided that humanity is treated to some extent as an end as well as a means. A more rigorous interpretation might appear to put all organizational decision making out of bounds. Certainly it would exclude involvement with the management of any conventional work organization in which people are routinely treated as instrumentalities for the achievement of externally defined goals.

There is, however, a class of situations that can pass even this stringent ethical filter. This is where work is carried out for and on behalf of organizations set up and run to advance or defend the interests of their members, where those members are disadvantaged and so are not candidates to become, as a result of the modeling intervention, oppressors of others.

Developed societies have such organizations in abundance. They take different forms, and certainly different names, in different countries. I am most familiar with Britain, where one can find such organizations as community centers, community health councils, parent–teacher associations, pressure groups advocating for the mentally ill, housing cooperatives, tenants' associations, production cooperatives, trade union branches, neighborhood environmental groups and residents' associations, community theater groups, and so on. Such groups confront the problems of making the most of their scarce resources in an environment characterized by complexity, uncertainty, and conflict.

The modeling profession has acted as if the only organizations in existence were those with hierarchical structures and large resources. Yet there are a plenitude of other organizations, concerned with their members' basic needs (health, education, shelter, employment, environment), that have decision problems no less recalcitrant. Indeed, a case can be made[8] that these community groups (as we may call them for short) have problems that are, if anything, more analytically and practically demanding than those of conventionally structured organizations. They must scheme and maneuver in a hostile environment with only marginal leverage to protect interests more vital than the next dividend. These community groups are the missing clientele of operational research. In terms of clues (á la Sherlock Holmes) to the mystery of the practice of modeling, they are the dog that didn't bark.

It is time to return to the words of Peter Szanton (of the New York City Rand Institute), cited earlier, and supply a fuller quotation. The complete passage reads, "Work such as the Institute's affects the balance of power as between government and governed. More and more clearly, our society is coming to regard interest groups — ethnic and racial associations, labor organizations, and neighborhood communities — as legitimate participants in local decision making. But these are groups with virtually no access to serious analytic support. I cannot say that our work has been much affected by this fact, but my own view is that this will be a problem of increasing social consequence".[9] Grassroots participation in decision making is in fact less in vogue today than Szanton anticipated nearly 20 years ago. However, legitimacy, ethics, and the practice of modeling are not best regarded as matters of fashion.

Szanton advocated the provision of such community groups with the analytic means of participating in public debate in a more informed and rational way. His expressed motivation was that the result would be an improvement in competence, relevance, and comprehensiveness in policy analysis. Others such as Cook[10] have been active in urging a wider social role for the profession in the interests of society and evidently of the profession itself. There are in fact a range of possible motivations all pointing in the same direction. They include improved social decision-making processes, an enhanced role for modelers, and a reduction in the disadvantage of the weakest groups in society.

This is not just pious speculation. Certainly in Britain at least, the operational research monolith is showing encouraging signs of developing a healthy diversity, and one of its most lively components is that of community operational research. This recent development is still weakly institutionalized relative to the longer established traditions of applied work for business, government agencies, and the military. However, it is appealing notably to the hearts, minds, and energies of the profession. Furthermore, its work constitutes an answer to many of the ethical doubts and conundrums that have been rehearsed in this chapter.

COMMUNITY OPERATIONAL RESEARCH

Community operational research has a prehistory that embraces Russ Ackoff's work in the Mantua ghetto in Philadelphia,[11] as well as the writings of Steve Cook[12] and others in Britain. However, its crystallization into an identifiable stream of practice can reasonably be dated to 1986. In that year, the (British) Operational Research Society adopted a proposal, made by myself as President of the Society, to support the establishment of a unit to carry out analytic work on behalf of disadvantaged groups.

Evidently such a decision itself has a history, based on a series of events and discussions in the British operational research profession over the previous 15 years. The details of this need not concern us here. What is relevant is the strength of support that the proposition attracted, culminating in a unanimous vote of the Society's Council.

When bids were invited, proposals to host the Community Operational Research Unit were received from 11 educational institutions, 9 of them universities. The breadth of this interest was encouraging, and still more so the revelation of the extent of work already being carried out for community groups. It seemed that the project of establishing a center for such work had surfaced a subterranean current of activities, formerly carried out in isolation and mutual ignorance.

Out of the 11 bids received, the one selected after a thorough, even painstaking, process was the proposal from Northern College in association with Sheffield City Polytechnic (since renamed Sheffield Hallam University). The Polytechnic has a strong operational research group teaching to master's level; Northern College is a residential adult education establishment outside Barnsley some 40 minutes drive away. Crucial to the success of their proposal was the regular use of the College by community groups, providing if not a captive then an available audience.

The Community Operational Research Unit became properly operational only at the start of 1989, with the appointment of its first full-time worker. His efforts have been supported by the work of both staff and students from Sheffield City Polytechnic and by outside consultants on particular projects. (By 1991 there were a further two half-time workers.) The first year's experience can give some flavor of the scope for community OR. It includes:

- work with a newly formed housing cooperative of some 350 units to help towards self-management of their estate;

- work with an association for the improvement of local maternity services to help the group plan its campaign to influence new birth facilities;

- work with a health and safety advisory center to help in a reorganization of its work practices and in evaluating and prioritizing its activities;

- work with a center for alternative technology on survey design, on queuing problems associated with a new development, and on, review of its cooperative decision-making methods and structures;

- assistance to a therapeutic community campaigning to prevent its closure.[13]

Progress to date has been heartening — in particular the demonstration that potential clients in profusion are eager to take advantage of these analytic services. Equally heartening, however, has been the reaction of the British OR community, for the Unit at Northern College does not now exist in isolation but is rapidly becoming the focus of a nationwide web of activity. A Centre for Community Operational Research has been established independently at the University of Hull. There are plans to establish a further unit in London. The Community Operational Research Network has sprung to life, to put practitioners who wish to carry out OR for community groups in their spare time in touch with potential clients. The Network has upwards of a hundred members, all full-time operational researchers in government, consulting, private industry and business, or academia. Local networks have been established in a number of areas to facilitate these contacts and to provide mutual support among practitioners embarking on this novel venture. It is not stretching language too far to refer to this activity as the community operational research *movement*.

As part of its decision to support the development of community operational research, the Operational Research Society identified the characteristics of its likely clientele. Such groups would

• exist to protect or advance the interests of their members,

• possess scant physical or financial resources,

• have no articulated management hierarchy, and

• operate internally through consensus or democracy.

While clients of community OR might offer goods or services for sale (as, for example, with production cooperatives), this would not be their principal purpose.

Working with such groups in a large part avoids the ethical dilemmas of "treating the humanity of others as a means" discussed earlier. The purpose of each organization is the advancement of the common interests that define membership, not the pursuit of some objective imposed externally. Such a group, relatively weak and disadvantaged, will not in general have the resources (nor, one might expect, the desire, opportunity, or priorities) to impose such treatment on other individuals or groups. If it did so desire, practitioners motivated to carry out community operational research would not likely be attracted to work with it.

These characteristics of the clients of community OR also have implications for the techniques and methods appropriate for working with them. For in dealing with the outside world, their relative weakness requires an ability to negotiate flexibly, to form alliances, and to maneuver skillfully. The crucial analytic need, therefore, is for assistance in structuring a problematic situation, not in solving a problem. Methods based on fixed objectives from which optimal resource allocations can be deduced are simply irrelevant. Indeed, what "resources"? The strength of such a group is only as great as the mobilization of its membership, and this depends on mutual trust and on consensus about policies. Methods must therefore be both participative, enabling free and diverse inputs to take place from individual members, and transparent, maintaining group ownership of the analysis rather than vesting it in experts who preside over an opaque mystery.

At this point, the phenomenon of community operational research intersects with another aspect of the fragmentation of the previously monolithic character of OR and related modeling activities. This is the development, worldwide in scope but particularly lively in Britain, of methods whose purpose is not the efficient solution of well-formulated problems but the provision of assistance in the process by which a number of stakeholders reach agreement on the nature of the problem that is of mutual concern. These "problem-structuring methods"[14] provide the type of participative and non-mystifying analysis that can in principle illuminate problem situations and facilitate consensus formation. They have indeed been significar. in the repertoire of methods deployed by the Community Operational Research Unit: the early projects listed above have involved the use of cognitive mapping, the Strategic Choice approach, and Strategic Assumption Surfacing and Testing. However, there has also been scope for more conventional approaches: financial modeling on spreadsheets (to explore policy options), survey design, and establishment of data-recording requirements. In the best empirical tradition of operational research, the approach must be to see which approach fits the situation.

The purpose of working with such groups is not the identification of some objectively "better" solution. Community operational research may indeed produce outcomes that are preferred by the client groups, although one can almost never know with certainty what would have transpired without the analytic intervention. However, the thrusts of the initiative are different: to counter the hegemony of one-sided modeling for those who need it least and to empower the weak by helping them to realize what is possible.

Community operational research is not unique in this respect. It bears many resemblances to work carried out by Scandinavian computer professionals and by Latin American social workers.[15] The phenomenon is in one respect unique, however, as the shouldering of a collective responsibility by the representative body of a professional grouping. As such it may be worthy of attention by similar organizations in countries other than Britain. The Operational Research Society would be more than flattered by imitation. A profession that embarks on an initiative of this kind should be prepared for the possibility of a bumpy ride. "Community," in the sense of non-hierarchical groupings concerned to protect their members' interests, is far from a conflict-free zone. There will undoubtedly be situations in which community operational research practitioners will find themselves in opposition to more conventional organizations that are, or could be, orthodox clients of mainstream operational research. Influential voices may be raised alleging that the alternative practitioners are bringing the profession into disrepute.

Such controversies will certainly be tests of commitment and resolve. I am confident, however, that the prophets of doom will be disregarded, and with good reason. No profession will do anything but enhance its standing by demonstrating its relevance to a wider range of social situations, by demonstrating its adherence to the ethical principle of equity, and by demonstrating its confidence in the central contribution it can make to improving the quality of social decision making. It is not only our alternative clientele who stand to benefit.

NOTES AND REFERENCES

1. Mulvey, J.M., Chapter 3 in this book.
2. This leaves out of account the special but significant case of models run on micros by the client without analytical assistance. But even these models do not write themselves.
3. Szanton, P.L. "Analysis and Urban Government: Experience of the New York City Rand Institute", *Policy Sciences*, 3, 153–161, 1972.
4. Rosenhead, J., "Operational Research at the Crossroads: Cecil Gordon and the Development of Post-war OR", *J. Opl Res. Soc.*, **40**, 3–28, 1989.
5. Rosenhead, J., "Custom and Practice", *J. Opl Res. Soc.*, 37, 335–343, 1986.
6. Ackoff, R.L., "Some Ideas on Education in the Management Sciences", *Management Science*, **17**, B2-4, 1970.
7. Churchman, C.W., "Operations Research as a Profession", *Management Science*, **17**, B37-53, 1970.
8. Rosenhead, J., "Custom and Practice", *op. cit.*
9. Szanton, P.L., *op.cit.*
10. Cook, S., "Operational Research, Social Well-being and the Zero Growth Concept", *Omega*, **1**, 647–667, 1973.
11. Ackoff, R.L., "A Black Ghetto's Research on a University", *Operations Research*, **18**, 761–771, 1970.
12. Cook, S., *op. cit.*
13. Community Operational Research Unit, Annual Report to September 1989, Northern College, Barnsley.
14. Rosenhead, J., (ed.), *Rational Analysis for a Problematic World: Problem Structuring Methods for Complexity, Uncertainty and Conflict*, Chichester: Wiley, 1989.
15. Sandberg, A. (ed.) *Computers Dividing Man and Work*, Stockholm: Arbetslivcentrum, 1979, and R. V. V. Vidal, "The Systematization of Practice", IMSOR, Technical University of Denmark, Lyngby, 1989 (mimeo).

Chapter 5

Where Do We Go From Here?

Part 1. Ethical Concerns and Ethical Answers

Saul I. Gass

IN THE BEGINNING

Like many of the readers of this book, I never had a formal course in ethics and ethical concerns. Whatever views I now have on good and evil, right and wrong, and duty and obligation are the result of interactions with family and friends, lessons learned and observed in school (especially during the early years), a couple of years in the Boy Scouts, and some time in the infantry during World War II. I never gave ethics much thought, but like most of us, I knew (intuitively?) what the correct ethical choice was when faced with a personal decision. Also, when I became involved in and/or observed the real world of business, politics, and human relationships, I developed a keen sense of smell and could filter out those things that were not kosher. I hardly gave any thought to what ethical framework applied to the professional aspects of my life. I "just knew" what was right and what was wrong.

For me as an operations research (OR) analyst, as a mathematician, and as a modeler, the ethical concerns of my profession first came out of the shadows and into the spotlight during the famous antiballistic missile (ABM) debate. In 1969, the Nixon Administration was considering the deployment of the SAFEGUARD ABM. A series of congressional hearings were held at which opponents and proponents of the ABM gave highly technical and contradictory testimony that relied greatly on what we would call military OR. The debate was heated and peppered with statements such as "... (so and so) makes an assumption that is plainly absurd".[1] One of the key participants in the debate, who was a member of the Operations Research Society of America (ORSA), wrote to the president of ORSA requesting "...ORSA to appoint a panel to consider some aspects of professional conduct during the

ABM debate..." and stated, "The question that I am raising is not whether the ABM decision was good or bad, but a question of professional standards and professional ethics".[2] Such a panel was established.

The reasons for citing the panel's work are twofold: (1) it marks the first time that many of us became concerned with codifying the professional and ethical aspects of the operations research profession, and (2) it produced a set of "Guidelines for Professional Practice".[3] Further, for many of us involved in governmental; business consulting and in the development of policy analysis models, the "Guidelines" addressed the issue of OR analysts as adversaries. I want to emphasize that the "Guidelines" are not an official part of the constitution or bylaws or "code of ethics" for ORSA. None exists at this writing! The reader can rightly ask, "Do modelers need a code of ethics?" My answer is a resounding "Yes!"

For the past few years, I have been asked to speak at the ORSA Doctoral Colloquium on "Ethics in the Practice of OR". The 30–40 student participants represent the "best and the brightest" of the profession's doctoral candidates. My presentation reviewed the need for a code of ethics and professional practices, discussed the "Guidelines" and other statements of correct professional conduct, and reviewed situations in which the OR analyst faces ethical concerns, especially in adversarial situations. As most of the Colloquium's students intended to become academics, the open discussion usually turned away from this more formal material to the ethical problems faced by doctoral students on their way to becoming new professors. Here the issues concerned professor–student relationships: (1) who gets credit and what is the order of authorship of journal papers, especially those that come out of the doctoral dissertation; (2) how one learns to obtain tenure and how objectivity can be brought to the process; (3) one's responsibilities as a peer reviewer of papers and proposals with respect to the confidentiality of the review and the reviewed material; and (4) one's responsibilities as a submitter of papers to be published to "play the game straight" with respect to unique submissions and the crediting of sources and material. Since much of my formal presentation at the Colloquium and concern with the ethics of the adversarial process has been reported elsewhere,[4] in this chapter I will discuss approaches to the development of a modeler's (OR analyst's) code of ethics in terms of what has been published by related professional societies and organizations. I will also discuss some of the student-related concerns mentioned above. However, before doing so, I want to stress what I believe to be the most important ethical concern that we as OR analysts face. That most important ethical concern is the adhering to proper professional practices in the development and reporting of our work.

MANAGING THE MODELING PROCESS

Most analysts conduct their work under the basic experimental, empirical, and objective procedures that guide all scientific inquiry. However, situations exist in which a model (usually computer-based) and its results are subject to debate because the problem environment is not clearly understood, as is the case for "greenhouse effects" models and for most modeling endeavors that attempt to simulate human behavior, such as "limits to growth" models. Since many of these models are used to guide governments and businesses in developing programs and allocating funds, it is clear to me that we must be able to scrutinize such models and their results in a traditional scientific manner. This means that data sources must be made available, methodological procedures must be reported in detail, and information must be provided so that external peer groups can replicate the results. This reasonable approach to what may be termed independent model evaluation is often rather difficult to accomplish due to the modelers not adhering to professional practices and standards, especially when it comes to proper and full documentation of the model, its computer implementation, and interpretation of the results. Although we cite "practices and standards," as noted above, I feel that providing for independent model evaluation is an ethical concern and appropriate for discussion here.

There are many "plausible and compelling" reasons for an analyst or project team to fall short of what you and I may feel is the ideal in modeling practice: the ideal is difficult to attain; resources (time, money, personnel) are always in short supply; the task is on a fast track and corners must be cut; no one is interested in having the model evaluated for proprietary or other reasons (it is a one-time affair); the modelers know they are right in their methodology; and so on. Certainly, not all modeling efforts have to be classics and not all models have to be or should be in a form for evaluation by others. But assuming that we want the results of our professional endeavors to be treated with due respect, all of our modeling activities must be accomplished according to accepted professional practices and standards. To this ethical end, we believe that the ORSA "Guidelines for Professional Practice" are most appropriate, especially when combined with "managing the modeling process" as given in the paper "Managing the Modeling Process: A Personal Perspective".[5] All analysts should strive to obtain the objects of these cited approaches, and all sponsors of such activities must recognize that doing the job correctly requires managerial and financial support that is often lacking in many of our analytical endeavors.

CODES OF ETHICS IN RELATED PROFESSIONS

As I have noted, ORSA, the professional OR society in the United States, does not have an official code of ethics. To me, this is an anachronism that should be corrected as soon as possible. I do not understand why such a situation exists, especially when most sister societies promulgate ethical codes. At best, I attribute this current state of affairs to inertia; at worst, I attribute it to hubris in the sense that ORSA feels itself to be "above it all" and believes that its members need no ethical guidance from it. Little effort would be required for ORSA to develop a code; an updating and tightening of the cited "Guidelines" would probably suffice.[6]

To soften my criticism of ORSA, I note that ORSA members have made efforts to develop a code of ethics, and ORSA has had Committees on Ethics and Professional Practice. I have copies of rather dated (1983) reports entitled "Code of Professional Standards" and "Guidelines for Professional Standards in Operations Research and Management Sciences" prepared by one such committee chaired by John D. Kettelle. I also note that there is no code of ethics for The Institute of Management Sciences (TIMS), the close sister society of ORSA. (About one-third of the members of ORSA and TIMS belong to both societies.)

It is not my purpose here and it would be out of place for me to propose a code of ethics for ORSA. However, for the interested analyst (and in the hope of renewing interest in a code of ethics for both ORSA and TIMS), I next cite and summarize appropriate codes from related societies and other sources that I feel offer appropriate guidance.

CODE OF PROFESSIONAL CONDUCT

I start our review of codes of ethics by citing the personal code of the statistician W. Edwards Deming.[7] It is truly an amazing document that carefully lays out (in seven sections and in 37 points) a code of professional conduct that is rigorous, strict, objective, and fair to both the client and the analyst. Deming's Code can serve as a paradigm for anyone who needs guidance on what a personal professional code should cover. To illustrate the careful thought that went into the writing of Deming's Code, I offer a very brief summary of his seven sections. In "Introductory Remarks", Deming states his aim in engaging in statistical consulting: "...to create new statistical methods, or to use existing methods (i) to help other scientists and professional men (sic) to improve their research; or (ii) to acquire new knowledge....; or (iii) to improve efficiency, uniformity, quality, service, and performance of product; or (iv) to achieve smoother operations and more effective administration and management in industry and government".[8] He then notes that he makes no solicitation for engagements but depends on recommendations.

In the sections "Obligations of the Client" and "Obligations of the Statistician", Deming spells out exactly what is expected from each. As a sampling plan is usually the basis of the statistical study, Deming states, in particular, that the "...final decision on the sampling procedure will be mine, and I will furnish instructions therefor in writing".[9] He then states that "The client will carry out my instructions for these experiments, and he will stand the expense therefor", and "The client will make no change in the statistical procedures without direction from me so long as my responsibility is in force".[10]

The fourth section deals with "Some Points Concerning the Interpretation of the Results, and about the Statistical Report or Testimony". Here Deming states the ground rules for his report and testimony, for example, "My report on the statistical reliability of the results, and any testimony, will be based on figures and other records that the client furnishes to me at my request for the result of the study".[11] And most important, he states: "I will not recommend to the client that he take any specific administrative action or policy as a result of the study. My responsibility ends with the statistical interpretation of the results".[12] Contrast this guideline with the objective of most operations research studies: to select and implement a solution (policy change).

The fifth section of Deming's Code deals with fees. Here he states: "My fees are not competitive. I do not tailor my participation to fit a price".[13] Section six's title is "There are not Proprietary Rights to Statistical Procedures". Here, Deming notes that "I may accept engagements from competitive firms.." and "...no client has proprietary right in any procedures or techniques that I prescribe".[14]

The final section of this "Code of Professional Conduct" includes, among other topics, the right to publish and present the theory and application of statistical procedures used in a study and states the basis of breaking the consulting engagement. It concludes with the following: "Any service agreed to places on me the obligation to resign at any time when in my judgment it appears that the study will not meet my requirements. I will issue an objection or a minority report at the conclusion of a study, if I am an advisor thereto but find that I can not concur with the stated limitations of published conclusions of importance".[15]

COMPOSITE CODES FROM THE SOCIAL SCIENCES

In his book *Ethical Dilemmas and Social Science Research*,[16] R. D. Reynolds gives two composite codes of ethics drawn from adopted codes of national associations and others, mainly from the fields of anthropology, political science, psychology, and sociology. Although these fields are, at best, peripheral to the practice of operations research, they do represent scientific fields that are based on human behavior and interactions, as does operations research.

The first composite code deals with "Investigators Responsible for Sponsored Research", while the second concerns "Social Scientists in Organizations". In stating the individual items of the composite codes, Reynolds indicates the number of societies that ascribe to each item. As this number can be treated as a surrogate for the importance of the item (the higher the number, the more important is the item), I next quote those items whose citation number was two or more (the number is given at the end of each item). The applicability of the quoted item to operations research should be clear (where appropriate, you need only substitute the term "operations researcher" for "investigator" or "social scientist").

Investigators Responsible for Sponsored Research — From 23 items, we have the following:

"Do not become involved in, or accept responsibility for, research which involves any unethical practice (4)".

"Do not agree to terms or conditions which would undermine the freedom and integrity of other scholars or researchers (2)".

"Be honest about limitations as an investigator or social scientist (2)".

"All conditions of relationship to the sponsor should be specific, explicit, and known to all members of the research staff (3)".

"Investigator should retain the right and responsibility for all ethical decisions related to a project (2)".

"Investigator(s) should retain the privilege of free and open publication whenever possible (2)".

"Investigator(s) should insist upon and ensure that full disclosure of sponsorship occurs in all reports (3)".[18]

Social Scientists in Organizations — From 29 items, we have the following:

"Interests and welfare of individual clients always take precedence over those of the patron, organization, and the like (4)".

"Employed social scientists must assure themselves that the files on individuals are maintained in a fashion to ensure confidentiality (3)".

"Employed social scientists should respect the rights and reputation of the employing institution (2)".

"Employed social scientists should respect the rights and ownership of the organization with regard to work produced for the organization (2)".

"Social scientists should insist on independence and autonomy in professional matters, even though related to organizational interests (4)".

"Employed social scientists should be responsible for determining their own limitations and capacity (7)".

"Social scientists, employed or in joint practice, should respect the professional independence of other social scientists (3)".

"In seeking employment, social scientists should be honest in stating their limitations and qualifications (5)".

"Social scientists in supervisory positions should ensure that descriptions of new positions for social scientists are presented honestly and completely (2)".

"Social scientists in positions of responsibility should ensure that any subordinate social scientist is encouraged to develop and advance professionally (2)".

"Social scientists in supervisory positions should never allow non-social scientists to conduct or take responsibility for social science activities or decisions (5)".[18]

A Code of Ethics for Policy Scientists

A code of ethics for policy scientists was proposed by Y. Dror.[19] The importance of this code was recognized by E. S. Quade, who has cited it in the many editions of his book *Analysis for Public Decisions*.[20] I next cite and offer in shortened form Dror's seven main ethical points. While Dror's points are directed toward analysts working in the public sector, I feel that all the issues raised are basic to all analysis studies, public or private.

(1) **Goals and values of the client** — if they contradict basic values of democracy and human rights, the analyst should not do the work.

(2) **Relationship between the analyst's values and the client's values** — if the client's beliefs contradict the basic values of the analyst, the analyst should resign.

(3) **Presentation of all viable alternatives** — the analyst should not hide an alternative because it contradicts his or her own personal values or preferences.

(4) **Increasing the judgment opportunities of the client** — the analyst should make a careful delineation of assumptions and presentation of unbiased sensitivity studies.

(5) **"Rubber stamping" a conclusion** — an analyst should refuse to prepare a study whose purpose is to support the client's conclusion that has been reached by other means.

(6) **Access to information** — the analyst should not work for a client who does not provide the necessary information and opportunities for presentation of the study and its findings.

(7) **Conflicts of interest** — the analyst must avoid all such conflicts, including the use of the study for personal gain and the making of recommendations on matters in which the analyst has a personal and private interest.[21]

National Institutes of Health

Over the past few years, some elements of the United States scientific establishment have been rocked by claims of fraud, data fabrication, and plagiarism. Such claims have occurred mainly in the medical and biological areas. Congressional committees have been formed to investigate specific situations. As OR/MS models and studies of public policy issues increase, our concern is with how we ensure that such work does not make the headlines in a negative way. Certainly, following the code of ethics for policy scientists and the ORSA "Guidelines" described above will help. In addition, I call your attention to the publication *Guidelines for The Conduct of Research at The National Institutes of Health*.[22] It states principles of behavior that all scientists are expected to follow. However, as these *Guidelines* note: "The formulation of these *Guidelines* is not meant to codify a set of rules, but to make explicit patterns of scientific practice that have been developed over many years and are followed by the vast majority of scientists, and to provide benchmarks when problems arise".[23] The *Guidelines* are cited and summarized as follows:

Supervision of trainees: The scientist, as a mentor and supervisor of a trainee, must avoid the involvement of trainees in research activities that do not provide meaningful training experiences but which are designed mainly to further research or development activities in which the mentor has a potential monetary or other compelling interest; training must impart to the trainee appropriate standards of scientific conduct, as conveyed by the mentor; and mentors have a responsibility to provide trainees with realistic appraisals of their performance.

Data management: The scientist must record research data and the results of research in a form that will allow continuous access for analysis and review; research data should always be immediately available to scientific collaborators and supervisors for review; research data should be retained for a sufficient period to allow analysis and repetition by others of published material from those data.

Publication practices: Timely publication of new and significant results is important, but fragmentary publications or multiple publications of the same or similar data are inappropriate; each paper should contain sufficient information for the informed reader to assess its validity with all the information that would enable scientific peers to repeat the experiment; tenure appointments and promotions should be based on scientific accomplishments and not on the number of publications.

Authorship: The privilege of authorship should be based on a significant contribution to the conceptualization, design, execution, and/or interpretation of the research study; each author should be willing to support the general conclusions of the study and be willing to defend the study.

Peer review and privileged information: Scientists have an obligation to participate in the peer review process (review of scientific paper, grant proposal, etc.); the reviewer should avoid any real or perceived conflict of interest; the review should be objective; all material under review is privileged information and should not be shared with anyone unless necessary to the review process.[24]

Ethics and the Consulting Engineer

In his book, *The Consulting Engineer*,[25] C. Maxwell Stanley cites the following Engineer's Pledge of Service adopted by the Iowa Engineering Society in 1942. We need only to replace "consulting engineer" with OR analyst and "engineering" with OR to make them applicable for our purposes.

1. "The consulting engineer will place service to mankind above personal gain and use engineering knowledge and skill to benefit humanity".

2. "The consulting engineer will render faithful, professional service to clients and honestly represent their interests".

3. "The consulting engineer will be governed by the highest standards of integrity, fair dealing, and courtesy in relations to others".

4. "The consulting engineer will encourage the development of the engineering profession and will contribute to improving the services of consulting engineers".[26]

Stanley also cites the following principles from the code of ethics of the American Consulting Engineers Council code of ethics. Consulting engineers:

1. "Hold paramount the safety, health, and welfare of the public in the performance of their professional duties".

2. "Perform services only in areas of their competence".

3. "Issue public statements only in an objective and truthful matter".

4. "Act in professional matters for each client as faithful agents and trustees, and avoid conflicts of interest".

5. "Build their professional reputation on the merit of their services and shall not compete unfairly with others".

6. "Associate only with reputable persons and organizations".

7. "Continue their professional development to their careers and shall provide opportunities for the professional development of those engineers under their supervision".[27]

The Association of Computing Machinery (ACM) Code of Ethics and Professional Conduct

The Ethics Task Force of ACM's Special Interest Group on Computers and Society published a draft code in the May 1992 issue of the *Communications of the ACM*.[28] This code is an update of ACM's previously adopted code (1972) and attempts to address new technological developments and other matters such as the impact of computer networks and computer viruses. The code is too long to cite here, but I note the following:

The ACM draft code is divided into four sections: "Section 1 — General Moral Imperatives" presents fundamental ethical considerations such as "Contribute to society and human well-being;" "Section 2 — More Specific Professional Responsibilities" addresses additional considerations of professional conduct such as "Strive to achieve the highest quality in both the process and products of professional work;" "Section 3 — Organizational Leadership Imperatives" relates to the individual as a leader and is concerned with such issues as "Create opportunities for members of the organization to learn the principles and limitations of computer systems;" and "Section 4 — Compliance with the Code" states how ACM members must view the code in terms of statements such as "Treat violations of this code as inconsistent with membership in the ACM". You are encouraged to read the full code given in ACM (1992).[28]

The Military Operations Research Society (MORS) Code of Ethics

MORS is a professional society, sponsored by the armed services and the Department of Defense, that "...seeks to enhance the quality and effectiveness of military operations research [MORS]". It has adopted the following Code

of Ethics and Professional Responsibilities for Practitioners of Military
Operations Research.

"Military OR Professionals must strive to be:

Honest, open, and trustworthy in all their relationships.

Reliable and consistent in the conduct of assignments and responsibilities,
always doing what is right rather than expedient.

Objective, constructive, and responsive in all work performed.

Truthful, complete and accurate in what they say and write.

Accountable for what they do and choose not to do.

Respectful of the work of others, giving due credit, and refraining from
criticism of them unless warranted.

Free from affiliation with others or with activities that would compromise
them, their employees, or the Society".[30]

Professional Conduct Guidance for the Fellowship for Operational Research

The Fellowship for Operational Research is a British professional society for
those engaged in operations research. The Fellowship has published a detailed
booklet entitled "Professional Conduct — Guidance for Fellows of
Operational Research",[31] which gives guidance in a host of matters such as
public responsibilities (e.g., dealing with television, radio, and newspaper
interviews), giving evidence for public bodies, giving testimony as an expert
witness, writing books and giving lectures, belonging to a trade union, and
being a consultant. Under "Ethical Obligations", the Professional Conduct
Guide notes the following:

"The obligations of an OR scientist will include:

1. Willingness to apply skills in an imaginative way, as distinct from
    merely obeying orders.

2. Recognition that the relationship with the employer calls for a high degree of mutual trust.

3. Willingness to work outside normal hours when the occasion demands.

4. A duty to carry out instructions from the immediate supervisor subject to the right to express reservations and have them recorded and to the right of access to the next higher level of authority.

5. Recognition that the consequence of sustained protest on a matter of conscience or professional judgement may be resignation or dismissal".[32]

The above descriptions, excerpts, and sources of professional standards are a rich mine of ideas for any society that wants to develop a code of ethics.[33] Further, it is clear that most professional societies have given much thought to the need of having an official code of ethics. Why ORSA does not as yet have a code of ethics is puzzling.

ETHICAL CONCERNS OF STUDENTS AND FACULTY

My participation in ORSA's Doctoral Colloquium has brought to the forefront a number of ethical concerns faced by students. As noted earlier, these concerns involve authorship of papers, obtaining tenure, the refereeing (peer review) of papers, submission of papers to journals, data availability and reproducibility of results, computational claims, duties as a referee or as an editor, professor–student papers, and professor–student relationships in and out of the classroom. I next comment on some of these issues.

From observation and my own experiences, a Ph.D. dissertation is often transformed into one or two (possibly three) publishable papers. Most Ph.D. advisors expect to have their name appear as one of the authors (usually the first author) on the first paper. This is accepted practice but is not fixed. It depends on how the advisor views his or her contribution, a view that may be at odds with the student's view. My suggestion is for the student not to fight this issue, for if the student felt that the advisor's contribution was minimal, at best, then something was wrong with the dissertation arrangement and that advisor should not have been involved in the first place. In such cases, given that there usually is more than one paper, the student should then be the sole author on the later papers. Much animosity can be generated between the advisor and student in such situations. The student needs to remember that the substantive guidance received from the advisor, even if it is only the advice, "Why don't you look at this area?" comes from the research base of the advisor and needs to be acknowledged. In turn, the advisor needs to recognize that the student has great pride in his or her accomplishments; the student wants the world to know of his or her discoveries. Also, the student is a fledgling researcher and has to be given confidence now that he or she is being tossed out of the nest.

Beyond the student–advisor situation, the more general question arises as to who should be listed as an author of a paper? B. J. Culliton describes criteria of authorship due to Arnold Relman, editor of the *New England Journal of Medicine*.[34] Although more appropriate for a natural science setting, Relman's definition of an author is of general interest. To qualify as an author, Relman suggests that a person must fulfill two of four requirements:

1. Conception of idea and design of experiment.

2. Actual execution of experiment; hands-on lab work.

3. Analysis and interpretation of data.

4. Actual writing of manuscript.

For operations research, the laboratory is replaced by the computer. I think most Ph.D. advisors would feel that they satisfy at least two of the criteria.

Culliton also addressed a related question of interest: Who owns the data? This should be of concern to operations researchers from two perspectives: fraud and reproducibility of experiments. Recent claims of fraud in the use of fabricated data have been made and proven in the biological and medical areas.[35] We know of no such situation in the operations research community, but who is to say that it does not exist. It has been suggested that a random audit of papers be made by journal editors![36]

Data availability becomes an issue when there is a need to reproduce an experiment, especially a computational one. Reproduction of results is essential to any scientific theory or claim. We all must document the data sources and data sets (even if they were randomly produced) so others may verify our results and be in a better position to compare new advances with old. The data and documentation should be available for proper scientific inquiry. There is, of course, the issue of proprietary data, especially when a claim is made by a private organization. However, if the claim is made in a scientific journal that required that the article containing the claim undergo peer review, then I do not see how such a peer review could have been done properly without access to the data.

The process of submission of papers to scientific journals and the peer review system is a mystery to most students and is laced with ethical problems. I offer the following comments. It is standard procedure to submit a paper for possible publication to only one journal at a time. If the paper is rejected by the first journal, it is proper to submit to a second, and so on. However, as the available referee pool is small, there is a good chance that you will get the same referee(s) and similar rejections. Some students have difficulty with the one-at-a-time submission process. They feel that if their paper turns out to be publishable, then the first one that comes through with a publication offer is the winner and is lucky to have it. These students have to recognize that the peer review process is completely voluntary and that referees, especially for a paper in a narrow field, are few and far between. Also, it is costly for the editor and associate editors who are also volunteers. Multiple submissions put a terrific strain on the peer-reviewed journal process; that is why most journals have a statement in their publication guidelines that the author agrees to a single submission. A word of advice. I recall reading a study (source lost) that showed that if a paper was rejected but the editor offered the author a chance to redo it based on the referees' reports, then the paper, if resubmitted, had a probability of 0.75 of eventually being accepted. Such a resubmission process can take more than two rounds and much time, however.

If you plan to publish papers, then you must also act as a referee when asked. There is no free lunch (the NIH guidelines cited above make the same point). As a referee, you must treat as confidential the ideas and results of the paper under review. If you offer the author a new and short proof for his or her great theorem that goes on for three pages, do not expect to become a coauthor and do not ask. The thanks to the anonymous referee is your payoff. Referee reports are anonymous, but you can ask the editor to make your name known to the author. There are good reasons for this, but most of us would not want that to happen.[37] The same confidential treatment applies to reviewing proposals for the National Science Foundation (NSF). Again, if you submit proposals to the NSF, expect to be called upon to be an NSF proposal reviewer. If your mind is set such that you cannot operate under these refereeing rules, then you should not be in a scientific field.

When refereeing papers, you need to avoid conflicts of interest. This is difficult, as you will get to know most of the other researchers in the field as professional and/or personal friends. There is no need to take yourself out of the refereeing loop when a friend's paper is sent to you. You just have to be extra careful in your evaluation and judgment to maintain both yours and the journal's integrity. Some journals will ask if there are any such conflicts but will not automatically disqualify you. The same is true for the NSF. You have to set the standard.

A final comment for students (and professors) is the following quote from Markie:

"I conclude, then, that the activity of friendship, for all its intrinsic value, is morally out of bounds for professors where actual and potential students are concerned. Instead of trying to be good friends to our students, we can and should use our energies to be good teachers to them".[38]

EXAMPLES OF ETHICAL SITUATIONS IN MODELING

I next offer some ethical situations that operations research analysts may encounter, as given by J. M. Mulvey:[39]

Let's get it done: due to cost, time, and ability, when developing and implementing a model, you do not or cannot consider all the factors.

Give me this answer: forcing a model to produce the user's desired outcome when such an outcome would not result if the model was used objectively.

Blow that whistle: not going to authorities after determining that your company's model and/or its use has been rigged to produce a desired outcome.

Incomplete exposure: using model-based results to state a position knowing that the position is sensitive to debatable hidden assumptions and is not really backed up by the data (which are uncertain) and that alternative positions could be justified using the same model.

I know I am objective: working as a consultant in an area in which you have a personal or corporate interest and not divulging such information to the individuals who must use your analysis.

I know what's best: not including an alternative because it may be better than the one you think the public should have.

Read my printouts: not describing the model and its results in a clear fashion and attempting to "snow" your audience by technical overkill.

Read my model: when using a model in an adversarial situation, not documenting the model and the analysis in a form that makes it readily available for evaluation and replication of results.

FINAL NOTE

It is essential for the future well-being of the operations research profession that its ethical concerns and problems be investigated and discussed in a more demanding fashion by its practitioners, academics, and related professional societies. The problems will not go away. They will become more prevalent with the wider use of operations research methodology. Operations research needs a code of ethics and professional practice.

NOTES AND REFERENCES

1. Caywood, T. E., Berger, H., Engel, J., Magee, J.F., Miser H.J. and Thrall, R.M., "Guidelines for the Practice of Operations Research", *Operations Research*, 19, 5, 1123–1258.
2. *Ibid.*, 1246–1247.
3. *Ibid.*, 1127–1137.
4. Gass, S. I. "Models at the OK Corral", *Interfaces*, **21**(6), 80–86, 1991 and S. I. Gass, "Public Sector Analysis and Operations Research/Management Science", in *Operations Research Handbook: Public Sector Analysis*, S. Pollock, M. Rothkopf, and A. Barnett, eds., NY: Elsevier, in press, 1993.
5. Gass, S. I. ed., "Managing the Modeling Process: A Personal Perspective", *European Journal of Operations Research*, **13**(1), 1–8, 1987.
6. Machol, R. E. "The ORSA Guidelines Report — A Retrospective, *Interfaces*, **12**(3), 20–28, 1982.
7. Deming, E.W., "Code of Professional Conduct", *International Statistical Review*, **40**(2), 215–219, 1972.
8. *Ibid.*, 215.
9. *Ibid.*, 215–216.
10. *Ibid.*, 216.
11. *Ibid.*, 217.
12. *Ibid.*, 218.
13. *Ibid.*, 218.
14. *Ibid.*, 219.
15. *Ibid.*, 219.
16. Reynolds, R. D. *Ethical Dilemmas and Social Science Research*, San Francisco:

Jossey-Bass Publishers, 1979.

17. *Ibid.*, 450–452.
18. *Ibid.*, 453–456.
19. Dror, Y. *Design for Policy Sciences*, NY: Elsevier, 1971.
20. Quade, E. S. *Analysis for Public Decisions*, 3rd edn, NY: North Holland, 1989.
21. Dror, Y. *op. cit.*, 219.
22. *Guidelines for the Conduct of Research at the National Institutes of Health*, National Institutes of Health, Bethesda, MD, 1990.
23. *Ibid.*, 3.
24. *Ibid.*, 4–8.
25. Stanley, C.M., *The Consulting Engineer*, NY: John Wiley, 1982.
26. *Ibid.*, 95.
27. *Ibid.*, 96.
28. ACM, "ACM Code of Ethics and Professional Conduct", *Communications of the ACM*, 35(5), 94–99, draft dated February 12, 1992.
29. *Ibid.*
30. MORS, *The World of MORS*, Military Operations Research Society, Alexandria, VA, undated.
31. Fellowship for Operational Research, *Professional Conduct — Guidance for Fellows of Operational Research*, London (undated).
32. *Ibid.*, 13.
33. Chalk, R., Frankel, M.S. and Chafee, S.B., "AAAS Professional Ethics Project - Professional Ethics Activities in the Scientific and Engineering Societies, AAAS, Washington, DC, 1981, and R. E. Machol, *op. cit.*
34. Culliton, B.J., "Authorship, Data Ownership Examined", *Science*, **242**, 658, 1989.
35. Roberts, L., "Misconduct: Caltech's Trial by Fire", *Science*, **253**, 1344–1347, 1991 and Hamilton, D.P., "NIH Finds Fraud in Cell Paper", *Science*, **2516**, 1552–1554, 1991.
36. See "News and Comments", *Science*, **242**, 657, November 4, 1988 and also Hamilton, D.P., "A Shaky Consensus on Misconduct", *Science*, **256**, 604–605, 1992.
37. Karp, A., "Are Signed Referee Reports the Answer?" *SIAM News*, 6, March 1992. Karp signs his referee reports!
38. Markie, P.J., "Professors, Students, and Friendship", p. 147 in Chapter 8 of *Morality, Responsibility, and the University*, S.M. Cahn, ed., Philadelphia, PA, Temple University Press, 1990.
39. Mulvey, J.M., Unpublished case studies and problems in ethics for OR analysts, Princeton University, Princeton, NJ, 1982.
40. Brewer, G.D., *Politicians, Bureaucrats, and the Consultant*, NY: Basic Books, 1973.
41. Brewer, G.D., "Operational Social Systems Modeling: Pitfalls and Perspectives", *Policy Sciences*, **10**, 157–169, 1978–79.

42. Brill, Jr, E.D., "The Use of Optimization Models in Public-Sector Planning", *Management Science*, **25**, 413–422, 1979.
43. Drake, A.W., Keeney, R.L. and Morse, P.M., *Analysis of Public Systems*, Cambridge, MA: The MIT Press, 1972.
44. Gass, S.I. and Sisson, R.L., eds., *A Guide to Models in Governmental Planning and Operations*, Potomac, MD: Sauger Books, 1975.
45. Goeller, B.F., "A Framework for Evaluating Success in Systems Analysis", P-7454, The Rand Corp, Santa Monica, CA, 1988.
46. Greenberger, M., "A Way of Thinking About Models", *Interfaces*, **10**(2), 91–96, 1980.
47. Greenberger, M., Crenson, M.A. and Crissey, B.L., *Models in the Policy Process*, NY: Russell Sage Foundation, 1976.
48. House, P.W. and McLeod, J., *Large-Scale Models for Policy Evaluation*, NY: John Wiley, 1977.
49. Kraemer, K.L. and King, J.L., "Computer-Based Models for Policy Making: Uses and Impacts in the U.S. Federal Government", *Operations Research*, **34**(4), 501–512, 1986.
50. Majone, G. and Quade, E.S., eds., *Pitfalls of Analysis*, NY: John Wiley, 1980.
51. Miller, L., Fisher, G., Walker, W. and Wolf, C., "Operations Research and Policy Analysis at RAND, 1968-1988", *OR/MS Today*, 20–25, December, 1988.
52. Mulvey, J.M., "Models in the Public Sector: Success, Failure and Ethical Behavior", Report SOR-89-19, School of Engineering and Applied Science, Princeton University, Princeton, NJ, 1989.

Chapter 5

Part 2. Responsible Policy Modeling

Warren E. Walker

> *For the analyst to be credible, moral claims to truth telling must be made,*
> *for without these, why should anyone ever believe what a planner or policy*
> *analyst says.*[1]

INTRODUCTION

The word "ethics" is based on a Greek concept of "trying to do good".
Unethical behavior means consciously doing something you know (or society
says) should not be done. These "things" include deception, bias, lying,
falsification, distortion, and withholding information. The issues discussed in
this book and the related issues that modelers (and, more generally, policy
analysts) confront in their work are usually not these "things". I maintain that
most of the issues discussed in this book are really not ethical questions but
are questions of good practice. The problems that are raised can generally be
resolved by following tenets of good professional practice. Those tenets can
and should be laid down in a "code of good practice"; they can and should
be taught to modelers-to-be in courses and to existing modelers in
professional journals. In summary, the question of "ethics in modeling" is
really a question of quality control.

Although by its very nature our profession has been unable to define
"good work" (modeling and policy analysis are crafts not sciences), we use
the scientific method in performing our work. This means that there are clear
principles of good practice that should be followed and clear yardsticks of
quality to judge analytic success. Among the most important of these (some
of which are made more specific later on) are:

- The work is open and explicit, and its results are verifiable and
  reproducible. Another analyst should be able to retrace the steps we have

226

used and obtain the same results. Our data, calculations, assumptions, and judgments should all be documented so that they can be subjected to checking, testing, criticism, debate, discussion, and possible refutation.

• The work is objective. Unlike other professions, the credibility of the work is not based on the renown of the analyst or his status in the profession. Its credibility is established by logical and empirical methods; its hypotheses are tested and verified. It strives to avoid the analyst's bias.

• Quantitative aspects of the work are treated quantitatively. Subjective judgments should be used as little as possible; when used, they should be noted explicitly.

In this chapter, I discuss the role of the modeler and policy analyst vis-a-vis the decision maker and describe the tenets of good practice that I believe should guide the modeler/policy analyst in carrying out his or her role. I assume throughout that the modeler/analyst is serving as a decision aider to a decision maker but is not the decision maker. There is an issue of corporate or public policy being discussed on which the modeler/analyst has been asked to shed light. This role requires providing the decision maker with the most accurate, complete, and unbiased information possible about the issue. It requires analysis, not advocacy. It requires clarity, not distortion. It requires a continual search for truth, not participation in partisan political debates.

I understand that there are many possible roles that a modeler may play in this world. Each role requires different types of behavior on his or her part. For example, the modeler as *advocate* will have a point of view on an issue and will use his or her expertise in modeling to support that position (much as a lawyer does to support a client). The modeler as *analyst* has a responsibility to provide unbiased information to all sides involved in an issue, regardless of who wins and who loses. When playing this role, the modeler must make sure that the models are as objective and value-free as possible. The analysis should be neutral; the interpretation of the results should be left to others who are playing different roles in the process. Thus, as Mood concludes, "the analyst is in the happy ethical position of being able to serve his sponsor best by adhering as rigorously as possible to the standard of complete objectivity and realism in assessing the effects of alternatives and designing an effective compromise".[2]

Both of the above roles are legitimate for the modeler. The ethical issue (the only truly ethical issue discussed in this chapter) is the responsibility on the part of the modeler to make it clear which of these roles he or she is playing at any given time. As Warwick and Pettigrew put it, "Policy researchers, like other citizens, have every right to express their policy opinions forcefully. But, on the other hand, these views need to be sharply separated from their research findings and their role as 'experts and social scientists'. That is, the modeler may switch roles but must maintain clarity about which role is being played whenever he or she 'performs'".[3]

In this chapter, I describe the primary elements of good practice that characterize the role of the modeler as analyst. The chapter is organized according to the stages generally executed in the course of a policy analysis study. For each stage, I describe the responsible and prudent behavior that should be expected of the modeler/analyst.

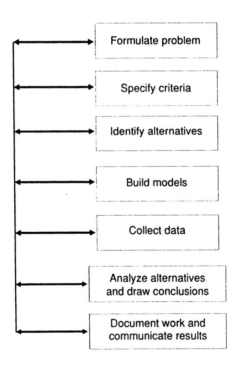

Fig. 1. The stages of policy analysis.

Although a study should generally include all of these steps, they will not necessarily be executed sequentially. Also, there will be a great deal of feedback and iteration. The stages (which are shown graphically in Fig. 1) are:

- Formulating (and reformulating) the problem

- Specifying criteria

- Identifying alternatives

- Building models

- Collecting data

- Analyzing alternatives and drawing conclusions

- Documenting work and communicating results.

GOOD PRACTICE IN EACH STAGE OF ANALYSIS

Formulating The Problem

The way a problem is formulated can easily bias the results. Therefore, care in problem identification and formulation is crucial. As we shall see, a major impediment to following good practice in carrying out a study has to do with resource constraints. Therefore, in identifying the problem it is important to be very specific and to limit the problem to a manageable size. To increase the chances for producing a useful product, the probable solution to the problem should lie within the jurisdiction of the client. As Wildavsky has observed, "Analysis requires creative juxtaposition of resources and objectives until problems are found that decision makers can solve — with the variables under their control, and within the time period available".[4]

Problem formulation should be viewed as a mini-study in its own right. It is dangerous to accept the client's problem statement as the problem to be addressed. It is also dangerous to accept the analyst's initial assessment of the problem. In many cases, it will be worthwhile to carry out some data collection and analysis to make an independent assessment of the problem situation and to translate it into a clear, focused problem statement. The analyst should understand such aspects as how the problem arose, why it is considered a problem, and what appears to be its cause. Inevitably, the problem statement will change over the course of the study. This should be planned for and welcomed.

In defining the problem, the analyst should strive to avoid what Warwick and Pettigrew call "conceptualization bias".[5] This means not defining the problem in a way that implies the conclusion (e.g., "the problem with our police department is that there are too few policemen") or in a way that implies values or judgments (e.g., defining the problem as "white flight" or "racial gap in achievement" may prejudge the cause as racial and interject a conservative bias). In the last example, one could equally well investigate differences among racial groups in achievement levels.

Specifying Criteria

In order to generate "good" solutions to the problem and to be able to choose among them, it is necessary to determine the desired objectives to be achieved. In most decision-making situations there are many stakeholders with many (often competing) objectives. A complete and unbiased analysis should take all of these differing objectives into account from the beginning. Failure to do so may result in finding solutions that may be far from the most desirable or may create strong opposition before they can be implemented.

One way to do this is to use the following three-step process:

(1) Identify all stakeholders (organizations, groups, or individuals) who stand to be significantly affected (positively or negatively) by the solution to the problem.

(2) Identify the objectives of each stakeholder. (Keeney[6] describes a structured approach for doing this.)

(3) Develop a comprehensive set of performance measures (quantitative and qualitative) that can be used to assess the effect of any solution on each objective. It is important to include the positive and negative effects on all stakeholders. For example, as Warwick and Pettigrew point out, "A study of the effects of shift work would have to be considered biased if it focused only on such harmful effects as ulcers and marital difficulties and neglected the potential advantages of night-time hours".[7]

One area in which this sequence of steps is generally not followed is in the area of environmental impact assessment (EIA). In most cases, the EIA process is to decide on a project (using the objectives and criteria of a small set of stakeholders) and do a *post hoc* impact assessment of that single alternative. Manheim asks: "How often are the preparations of impact assessments separate from, and after, the major technical effort devoted to development of alternatives? How often is such serious and thorough attention given to social and environmental analyses that the basic conception of a project changes significantly? Or are these analyses simply additional pieces of paper that create an appearance of attention without changing conclusions?"[8]

Identifying Alternatives

As Walker[9] explains, it is important to specify a wide range of alternatives for consideration and to include as many as stand any chance at all of being worthwhile. If an alternative is not included in this step, it will probably never be examined. If it is not examined, there is no way of knowing just how good — or bad — it may be. The best policy may not be chosen by the analysis because it was never considered as an alternative. Alternatives should not be excluded merely because they seem impractical or run contrary to past practice or conventional wisdom. Personal judgments on such issues should be withheld. The analysis will show whether the benefits to be derived outweigh the costs of making such radical changes. The importance of the need to withhold personal judgments and withstand political pressures (and the difficulty of doing so) cannot be overemphasized. Miser and Quade explain this dilemma:

> "There are many forces that tend to restrict the range of alternatives likely to be examined. Some of the strongest are biases of various sorts due to the unconscious adherence to an organization's "party line" or cherished beliefs or even mere loyalty. It can also happen that the analyst, in talking with the decision maker or his staff, becomes aware that the decision maker (or his superior) doesn't like certain kinds of alternatives. He may sense that it is both useless and hazardous to even give the impression that he might advocate these alternatives as possible solutions. As a result, the development of such alternatives is likely to be neglected or forgotten, thus leading to inferior results".[10]

Alexander shows how the development and review of options for the United States' Vietnam policy between 1961 and 1968 was restricted by exactly such limitations. "It is striking how few, and how similar, the options were which emerged from the review process and its various iterations...The policy establishment repeatedly closed ranks against ideas suggested by what they perceived as a hostile environment, or proposed by establishment mavericks".[11] Similarly, in the field of transportation planning, Manheim asks, "In how many highway studies are mass transit and/or not-meeting-future-demand alternatives seriously considered? In how many airport studies are alternative modes or not-meeting-future-demand seriously considered?"[12]

Building Models

Once the alternative policies have been selected, each one has to be examined to determine the likely consequences of its implementation in terms of the performance measures specified in the previous step. This is usually done using a model of the real world (since trying the policies in the real world is likely to be impossible, impractical, expensive, or dangerous).

We are all aware that models are not reality but only approximations or representations of reality. We always make judgments and compromises in developing practical models. Because of the high degree of judgment that goes into model development, there is a need for careful scrutiny in their building and use.

Principles that should be followed in building, verifying, validating, and documenting models have been discussed throughout this book. One of the most important of these is that the model should be understandable to others (i.e., it should not be a "black box"). Its underlying assumptions must be clearly stated, so that the user can verify its appropriateness to the situation at hand. More important (and difficult), the modeler should specify the situations for which the model is applicable and for which it will produce outputs that can be believed. Use of the model in other situations is bad practice. It is very rare that a model can be validated according to the meaning of the term in the physical sciences — i.e., finding that it agrees with reality. The best that can normally be done is to check its validity according to the definition given by Miser and Quade who state that validation is "the process by which the analyst assures himself and others that a model is a representation of the phenomena being modeled that is *adequate for the purposes* of the study of which it is a part".[13] It is also possible to use a model that cannot be validated in any fully adequate sense (such as by comparing it with actual data) for certain legitimate purposes. Hodges discusses the "six (or so) things you can do"[14] with such a model. (For example, you can use it as a bookkeeping device, as an aid to thinking and hypothesizing, or as an aid in selling an idea which is illustrated by the model.)

The modeler should also be sensitive to the fact that the type of model chosen for an analysis, or the way that model is constructed, will, in a large part, determine the conclusions. For example, the constraints in a linear programming model will automatically exclude many solutions. Warwick and Pettigrew suggest a similar caution in using econometric models: "Large-scale, often mathematically elegant econometric models...are shaped by their base assumptions, and these assumptions, though often unstated and untestable, significantly mold the policy conclusions that are drawn from such models".[15] It is the modeler's job to make sure that no reasonable solutions are excluded. Greenberger, Crenson, and Crissey[16] suggest that for large, important policy models, "counter modeling" teams be created and funded that would have the responsibility for critically appraising the model from different value perspectives. This might include building and exercising alternative models, as well as performing sensitivity experiments on the model structure and parameter values.

A more extended discussion of various aspects of good practice in modeling is given by Allen *et al.*[17]

Collecting Data

Most of us believe the GIGO principle of modeling — garbage in, garbage out. But many do not pay attention to its implications for good practice. Many of these implications are laid out in the 1971 *Guidelines for the Practice of Operations Research.*[18] In particular, they recommend the following:

- **General.** "Obtain access to all information that can reasonably be thought to be needed for the problem's solution, or to have a possibly significant bearing on it".

- **Conducting a Study.** "Assemble relevant information and data of verified reliability, or if not available, inputs of judiciously and suitably estimated unreliability, so that the impacts of uncertainty can be assessed in the results".

- **Reporting a Study.** "Report clearly...the essentials of the input information (and inaccuracies therein)...Complete exposition of assumptions, methodologies, objectives, basic data, and conclusions should be provided in scientific works to permit replication of the procedures described by objective observers".

I have little to add to this advice. However, I would emphasize the need (and the difficulty) to understand what the data mean and to assess their accuracy. Cleaning and validating the data to make sure that they are sensible are of utmost importance. Editing the data to detect errors before any analysis will save an enormous amount of effort later in the study and will avoid drawing erroneous conclusions.

Yet, gathering and cleaning data and building good databases takes a great deal of time and resources, and there is rarely enough time or resources budgeted for these activities. Tufte presents the following rule of thumb for deciding how much effort is required: "(1) add up all the time for *everything* you can think of — editing the data, checking for errors, calculating various statistics, thinking about the results, going back to the data to try out a new idea, (2) then multiply the estimate obtained in the first step by five".[19]

Analyzing Alternatives and Drawing Conclusions

For the unsuspecting analyst, this step in the analytical process has the greatest potential for bias. It also is the easiest step in which to avoid bias if the analyst is careful to maintain the distinctions described at the beginning of this chapter between (1) decision aiding and decision making, and (2) between analysis and advocacy. This means that the analyst must strive to present objective and complete information to all stakeholders and leave interpretation to others. It does not mean that he or she cannot point out interesting results; but it does mean that he or she should avoid playing the role of decision maker or advocate. (For example, the analyst may say, "if you have this objective function, you would favor this alternative", but would usually not say "here is the best policy".)

A problem in this stage is how to synthesize the impacts and present them to the decision maker(s) without imposing the analyst's values. Many approaches (e.g., cost/benefit analysis) weigh each impact by its relative importance and combine them into a single number representing money, worth or utility. This number is then used to compare the alternatives. However, the measure of worth depends strongly on the weights given to the different impacts. If the analyst decides on the weights, he or she is imposing his or her values.

An alternative approach that avoids this problem is to present the impacts in the form of a matrix called a *scorecard*, which presents a column of impacts for each alternative, with each impact expressed in units natural to it. Figure 2 is a sample scorecard that presents selected results from a RAND project that compared three alternative ways to protect a Dutch estuary from flooding.[20] The scorecard has several advantages. It presents a wide range of impacts and permits a decision maker to give each impact whatever weight he or she deems appropriate. It helps the decision maker to see the comparative strengths and weaknesses of various alternatives, to consider impacts that cannot be expressed in numerical terms, and to change his or her subjective weightings and note the effect this would have on the choice. When there are multiple stakeholders, the scorecard has the additional advantage of presenting the information needed for debating the tradeoffs and negotiating the choice among the alternatives without requiring explicit agreement on weights for different social values. (It will generally be easier for different stakeholders to agree on the alternative they prefer than on the weights to assign to the various impacts.)

	Closed case	SSB case	Open case
	Alternatives		
SECURITY			
Land flooded (ha) in 1/4000 storm (90% prob.)	0	0	400
Technical uncertainty	None	Scour	Dikes
Expected land flooded during transition pd. (ha)	430	200	530
RECREATION			
Added shoreline (km)	17	11	6
Added sea beach visits (1000/yr)	338	0	0
Decrease in attractiveness of area	None	Minor	Major
Major tourist site created?	No	Yes	No
Decrease in salt-water fish quantity (%)	75	0	25
NATIONAL ECONOMY (PEAK YEAR)			
Jobs	5800	9000	5700
Imports (DFL million)	110	200	130
Production (DFL million)	580	940	560

Rankings: ▨ Best ╲╲╲╲╲ Intermediate ☒☒☒☒ Worst

Fig. 2. Sample scorecard from policy analysis of the Oosterschelde (POLANO) project.

Documenting Work and Communicating Results

Modelers and policy analysts try to use the scientific method whenever possible, which means that they must document their work. As Miser and Quade put it, "Documentation is as much a part of the professionalism of systems analysis as it is of pure science, and the need to have full and clear records at the end of the project should be recognized and responded to. [One should, therefore,] do what is necessary at each step of the analysis to build the records that will allow others to see clearly what was done, and, if they should ever desire, to duplicate or extend the work".[21]

The ORSA Guidelines,[22] after stating that "the ultimate effectiveness of a study critically depends on how well its findings are communicated, understood, and then acted upon", suggests that the analyst should:

- Insofar as possible, use the vocabulary of his client, introducing only such new concepts and terminology as are essential to understanding the findings...

- Report clearly the problem formulation finally adopted..., the key assumptions used, the major alternatives considered, the essentials of the input information (and inaccuracies therein), the criteria employed..., the findings (including their sensitivity to realistic changes in assumptions, or the uncertainty in data), and their implications for policy and action.

- Delineate conscientiously what was accomplished by the study, and perhaps even more important, what was not considered or accomplished.

- Specify limitations on methodology or conclusions that should be observed, and spell out with candor instances where definitive results are not provided by the analysis.

- Set the study in the larger context appropriate for it.

- Prepare a written report on at least two levels: one for the client...and another fully technical report that can be examined by operations — and systems — research scientists.

- Be prepared to participate in any follow-up or implementation activities, both to assist with them and to evaluate their results".

As in the previous steps, doing a high-quality, professional job of documenting the analysis and communicating the results requires a lot of time and effort. As Holling reports, "Our experience is that at least as much effort must go into communications as goes into analysis".[23]

A STRATEGY FOR ENCOURAGING RESPONSIBLE MODELING

In the context outlined above, ethical modeling can be equated with quality modeling. In any profession there will be unethical persons—persons who willfully and consciously choose the path of unprofessional conduct. This book is not addressed to such persons. It is, instead, addressed to persons who have the best of intentions but might err because they are unaware of or insensitive to some aspects of good practice. The authors of chapters in the book have tried to shed light on some of these.

What would be the best way for us to get the word out more broadly to the world of modelers and would-be modelers? My answer consists of two parts: (1) develop a clear statement of good practice that reflects the current contexts within which modeling takes place and the current state of information technology and (2) explain this view of good practice to current and future modelers. To accomplish these objectives, I recommend two complementary efforts: (1) update the 1971 ORSA "Guidelines for the Practice of Operations Research" and disseminate them widely and (2) prepare associated teaching materials and campaign for their use in college-level modeling curricula.

Why not solve the problem of "quality in modeling" by having modelers certified, as has been suggested by some in our profession? My answer is threefold:

1. *Identification of modelers is impossible.* There is no well-defined group of persons that can be identified as "modelers". Twenty years ago, there was a professional community of modelers. Nowadays, anyone with access to a computer can become an instant modeler.

2. *Certification of modelers is unreliable.* There is no way to police such a diverse group of people to assure that they are doing the "right thing". There is no reason to believe that potential clients can be (or should be) encouraged to use only certified modelers.

3. *Certification is inefficient.* Implementing a certification program would require setting up a bureaucracy to define criteria, develop testing instruments, administer and grade exams, etc.

On the other hand, it should not be too difficult to update ORSA's *Guidelines*. Most of what is already in the *Guidelines* has stood the test of time and is as applicable in today's world as it was in 1971. I would like to see the guidelines expanded in two directions:

1. To include a broader perspective on the role of modeling in public-sector and private-sector policy analysis. I have outlined some of the features of good practice in policy analysis in the previous section of this chapter.

2. To include guidelines for good practice in the building, implementation, and use of both embedded computerized decision procedures (CDPs) and CDPs transferred from modelers to decision makers for their use (and possible abuse). I discuss these two relatively new categories of models below.

Embedded CDPs are simple, repetitive decision algorithms that are built into operating systems. They make and implement their decisions without any human intervention. They generally function as automatic control devices. One example of an embedded CDP is the climate control system in a modern office building, which regulates the flow of air (hot and cold) throughout the building. Such systems are usually designed to optimize some objective function but not necessarily the user's objective function. As such, they take control away from the individual, who is no longer the decision maker. (For example, climate control systems are usually designed to minimize the building's energy consumption, not necessarily to maximize human comfort.)

CDPs that are transferred from modelers to decision makers were discussed by Mulvey in an earlier chapter of this book.[24] These are mathematical models or heuristic procedures that have become routine parts of an organization's standard procedures for making decisions. The models are generally large and complex. For example, Mulvey cites the FAA's flow-control CDP, which "decides" which planes on the ground at airports throughout the United States will be delayed and by how much, based on its forecasts of future traffic at the country's busiest airports.

The main problem with both of these types of models is that, although the models themselves may have been built and tested by persons who followed the traditional rules of good practice, the original builders and users have lost control, and the models have assumed a life of their own. In the past, most modelers remained closely connected to their models and their users. Now, this is not the case. Most of the time the modeler is unknown or the user is unaware that he or she is having decisions made by a model.

The nature of CDPs require additions to the traditional rules of good practice. Some (far complete) suggestions would be:

- Before implementation or transfer, verify and validate the model inside and out, under as many conceivable scenarios as possible to make sure that it will not do something ridiculous. You can be sure that the "worst case" scenario will occur sometime, somewhere.

- Try to remain in touch with a CDP until it is "hard-wired" into an operational procedure to try to ensure that it will be used for only those situations for which it was tested.

- When transferring a model to a client, make clear the situations in which it can be used with confidence and (perhaps more important) the dangers of using it in other situations.

- Try to get the design and implementation of the model to be reviewed periodically to determine if it is still appropriate to use in the ways it is currently being used.

- Promote the insertion of a human decision maker in any decision-making loop in which the dangers of a "fatal error" from automatic decision making are great (which may precipitate outside pressure for regulation of CDPs).

Once the ORSA *Guidelines* have been revised, updated, and expanded, they must be disseminated throughout the modeling community. To accomplish this, I propose a two-track dissemination and education process, one for current modelers and one for modelers-to-be.

There are many mechanisms for bringing the new *Guidelines* to the attention of modelers. There is no reason to restrict attention to any one. All should be employed. The possibilities include publication in a special issue of *Operations Research*, publication in other journals and magazines read by modelers, and presentation at professional meetings.

More difficult will be the education of modelers-to-be. As already mentioned, modeling is no longer restricted to a small club. Modelers may receive their academic training in a large variety of disciplines. There is a need to sensitize educators in these disciplines to the dangers of poor practice and introduce them to the new *Guidelines*. The more attention and interest we can generate in the new *Guidelines*, the more likely it is that educators in other disciplines will hear about them and give them some thought. The hope is that there will be increased recognition among all types of modelers that we are all in this boat together and that our professional careers depend on maintaining a high regard for models and their use in decision making.

NOTES AND REFERENCES

1. Forester, J., "What Analysts Do", Chapter 4, 55, in W.N. Dunn (ed.), *Values, Ethics, and the Practice of Policy Analysis*, Lexington, MA: Lexington Books.
2. Mood, A. M., Introduction to Policy Analysis, North-Holland, New York, 286–287, 1983.
3. Warwick, D.P. and Pettigrew, T.F., "Toward Ethical Guidelines for Social Science Research in Public Policy", Chapter 14 in D. Callahan and B. Jennings (eds.), *Ethics, the Social Sciences and Policy Analysis*, New York: Plenum Press, 356, 1983.
4. Wildavsky, A., "Principles for a Graduate School of Public Policy", *Journal of Urban Analysis*, **14**(1), 3–28, 1977.
5. Warwick, D.P. and Pettigrew, T.F., *op. cit.*, 342.
6. Keeney, R.L., "Structuring Objectives for Problems of Public Interest", *Operations Research*, **36**, 396–405, 1988.
7. Warwick, D.P. and Pettigrew, T.F., *op. cit.*, 342.
8. Manheim, M.L., "Ethical Issues in Environmental Impact Assessment", *Environmental Impact Assessment Review*, **2**(4), 315–334, 1981.
9. Walker, W.E., "Generating and Screening Alternatives", Chapter 6 in H.J. Miser and E.S. Quade (eds.), *Handbook of Systems Analysis: Craft issues and Procedural Choices*, New York: Elsevier Science, 220, 1988.
10. Miser, H.J. and Quade, E.S., (eds.), *Handbook of Systems Analysis: Overview of Uses, Procedures, Applications, and Practice*, New York: Elsevier Science, 184, 1985.
11. Alexander, E.R., "The Design of Alternatives in Organizational Contexts: A Pilot Study", *Administrative Science Quarterly*, **24**, 382–404, 1979.
12. Manheim, M.L., *op. cit.*
13. Miser, H.J. and Quade, E.S., *op. cit.*, 301.
14. Hodges, J.S., "Six (or so) Things You Can Do with a Bad Model", *Operations Research*, **39**(3), 355–365, May–June 1991.
15. Warwick, D.P. and Pettigrew, T.F., *op.cit.*
16. Greenberger, M., M.A. Crenson, and B.L. Crissey, *Models in the Policy Process*, New York: Basic Books, 1976.
17. Allen, P., Bennett, B., Carrillo, M., Goeller, B. and Walker, W., "Quality in Policy Modeling", *Interfaces*, **22**(4), 70–85, July–August 1992.
18. ORSA Ad Hoc Committee on Professional Standards, "Guidelines for the Practice of Operations Research", *Operations Research*, **19**(5), 1123–1148, September 1971.

19. Tufte, E.R., *Data Analysis for Politics and Policy*, Englewood Cliffs, NJ: Prentice Hall, 1974.

20. Goeller, B.F. *et al.*, *Protecting an Estuary from Floods—A Policy Analysis of the Oosterschelde: Vol. I, Summary Report*, R-2121/1-NETH, The RAND Corporation, Santa Monica, 1977.

21. Miser, H.J. and Quade, E.S., *op. cit.*, 301.

22. ORSA Ad Hoc Committee, *op. cit.*

23. Holling, C.S., *Adaptive Environmental Assessment and Management*, Chichester, England: John Wiley, 120, 1978.

24. Mulvey, J.M., "Models in the Public Sector: Success, Failure and Ethical Behavior", a chapter in this volume, 1994.

Chapter 5

Part 3. Society's Role in the Ethics of Modeling

Edith H. Leet and William A. Wallace

As computers become both more sophisticated and more user-friendly, it seems inevitable that they will be used more widely and more often in modeling for decision making. But merely because a greater number of people are using computers and "know the right buttons to push", it does not follow that these people understand the technology and appreciate its limitations. On the contrary, the very "user friendliness" of modern computers is likely to lead to more people using them who have far less knowledge and understanding of how and why they work than early computer users did.

Power remains, therefore, in the hands of those who design the programs. And with that power comes the potential for misplaced trust and abuse. Our society is learning to "trust" computers, to see them as "objective" instruments that process data and give us "objective" answers to the questions we ask of them. As a society, we need continual education and reminders of the fact that computers are only as objective as the people who program them and that none of us, even with the best of intentions, can ever be perfectly objective.

Models, abstractions of reality, are developed to help summarize and make comprehensible the data collected and processed by information technology. These models are then embedded in software and become components in computer-based decision support systems.

Values are inherent in any model, whether they are incorporated intentionally or unintentionally. The potential thus exists for the results to be skewed. How much we allow them to be skewed depends in large measure on how well we understand how and why values can affect the model's results. We need to foster a healthy skepticism toward models that will allow us to use them and profit from what they can do for us, but not to trust them to make the decisions for us.

In a very real sense, most of us as model users are like the immigrants

242

who streamed through Ellis Island a century ago: We stand at the gate of a shining new society, full of potential, full of the promise of a better life. We are eager to see the fulfillment of its promises, but most of us don't know the language. We must rely on those who do know the language to interpret for us.

Like the immigrants of old, some of us may fall prey to unscrupulous interpreters, for deliberate deception and dishonesty are always a possibility. But they are not necessarily the greatest danger. More harm can be caused by unintentional means.

When model builders and model users fail to recognize the values and assumptions on which a model is based or fail to take into account all the groups who would be affected by a model's results, they set the stage for decisions that appear "scientific" and "objective" on the surface but that can have far different effects from those intended.

To mitigate against these dangers, we as a society must take a proactive role in the model-building enterprise. We must encourage model builders to set and adhere to high ethical standards. We must provide the laws and regulations to enforce those standards and also encourage continuing education about and awareness of the role of models in decision making, including both their benefits and their drawbacks.

Model building today is far from being a cohesive and self-regulating profession. Nor is it likely to become one quickly. The questions of values and ethics are just beginning to be addressed by established model builders, and anyone with appropriate computer know-how and entrepreneurial enthusiasm can enter the field at any time. It is apt to be many years before model builders will be able to agree on a code of ethics, let alone have any means of enforcing it within their ranks.

As a society, we cannot afford to wait for modelers to adopt and adhere to their own professional standards. We all have a vital interest in addressing the ethical issues and providing appropriate safeguards against abuse. The ramifications of decisions based on models can be too broad and far-reaching, both in terms of public policy and private investment, for society not to take an active role.

We recommend that the discipline of Applied Ethics be employed to provide the knowledge upon which we can begin to prescribe ethical behavior for model builders. Applied Ethics is an academic endeavor that examines moral behavior with the aid of philosophical theories. The goal of such examination is to sensitize practitioners to where ethical dilemmas arise in practice; to help them become aware of how their own values enter into decision making; to alert them to alternative courses based on alternative values; to stress where additional facts are needed for solution; and to encourage reflective decision making in order to arrive at a consensus that satisfies a "greater moral good".[1] Achievement of these goals would contribute greatly toward a modeler's code of ethics.

Several of the writers in this book have documented cases in which the values incorporated into a model led to inaccurate, questionable, or contested results. Policies and decisions based on those models were similarly called into question. In other cases, the objectivity of the models was not questioned by their users, and the writers thought it should be. Whether or not there was any deliberate intent to falsify or skew the models, they were found to be flawed, and thus the decisions based on their results were also flawed.

As a society, we also cannot afford to leave all the ethical decisions in the hands of the model builders and model users. Even the most ethical and public-spirited of them are still limited by their own perspectives. Their values and interests are bound to be different from those of other individuals and groups in society who may be affected by their decisions.

Our society has a responsibility to protect the rights of all of its citizens, and one way to do that is to ensure that their values and interests are taken into account when models are constructed. Companies must now make environmental impact studies to anticipate the effects of their plans on the physical environment. Society may do well to require that similar "human impact studies" be made before companies can proceed with their projects.

Finally, both modelers and society have a responsibility to increase public awareness of the nature and role of models in decision making. Model builders have a responsibility to inform their clients of what their models can and cannot do, of what questions their models can and cannot answer. Model users have a responsibility to inform themselves about models' potentials and limitations, to be intelligent consumers who demand from model builders models that reflect reality as faithfully as possible.

Society has a responsibility to educate its citizens about the nature of computers as decision aids. Recognition and discussion of the human values and ethical decisions inherent in computer-based decision making should begin when students are first introduced to computer technology. It should continue and increase in depth and complexity throughout a student's formal education, and it should be a fundamental part of professional debate and media awareness in adulthood. Although some of us may never learn to speak the language of modelers fluently, we all can learn Barabba's Law: "Never say the model says". With that understanding and healthy skepticism, we can begin to harness the power of modeling as a tool for human decision makers — not as the decision maker itself.

NOTES AND REFERENCES

1. Pinkus, R., *Personal Communication*, 1990.

APPENDIX A: Authors
(Affiliation at Time of Publication)

John Allison, Director, Center for Legal and Regulatory Studies, Department of Management Science and Information Systems, Graduate School of Business, The University of Texas at Austin, Austin, TX 78712-1175, USA.

Vincent P. Barabba, Executive Director, Department of Marketing and Planning, General Motors Corporation, 3044 West Grand Boulevard, Room 6-235, Detroit, MI 48202, USA.

Harold D. Carrier, Assistant Professor, School of Management, Rensselaer Polytechnic Institute, Troy, NY 12180-3590, USA.

Abraham Charnes — Deceased. Formerly, Director and Professor, Center for Cybernetic Studies, Department of Management Science and Information Systems, Graduate School of Business, The University of Texas at Austin, CBA 5.202, Austin, TX 78712-1170, USA.

William W. Cooper, Professor of Management, Accounting, Management Science and Information Systems, Graduate School of Business, The University of Texas at Austin, Austin, TX 78712-1170, USA.

Stephen E. Fienberg, Office of the Vice President (Academic Affairs), York University, 4700 Keele Street, North York, Ontario, Canada, M3J IP3.

Saul I. Gass, Professor of Business Management, College of Business and Management, University of Maryland, Tydings Hall, Room 0141B, College Park, MD 20742-7215, USA.

Suzanne Harris, Stratton Associates, Inc., Washington, DC, USA.

Edith H. Leet, EHL Editorial Services, 951 Myrtle Avenue, Albany, NY 12203-1817, USA.

John D. C. Little, Maverick Bunker Professor, Sloan School, Institute Professor, Operations Research Center, Massachusetts Institute of Technology, Cambridge, MA 02319, USA.

Richard O. Mason, Carr P. Collins Distinguished Professor, Department of Management Information Systems, Cox School of Business, Southern Methodist University, Dallas, TX 75275, USA.

Paul D. McNelis, SJ, Associate Professor, Department of Economics, Georgetown University, Washington, DC 20057, USA.

John M. Mulvey, Professor and Director of Engineering Management Systems, Department of Civil Engineering and Operations Research, Princeton University, Princeton, NJ 08655, USA.

Jonathan Rosenhead, Professor, Operations Research Department, The London School of Economics, Houghton Street, London WC2A 2AE, UK.

N. Phillip Ross, Chief of Statistical Policy Branch, Office of Standard Regulations, Department of Policy Planning and Evaluation, United States Environmental Protection Agency, Mail Stop P.M. 223, 401 N. Street South West, Washington, DC 20460, USA.

Toshiyuki Sueyoshi, Assistant Professor, School of Public Policy and Management, Ohio State University, 1775 College Road, Columbus, OH 43210-1399, USA.

Warren Walker, Senior Policy Analyst, Systems Sciences Department, The RAND Corporation, 1700 Main Street, Santa Monica, CA 90406-2138, USA.

William A. Wallace, Professor, Department of Decision Sciences and Engineering Systems, Rensselaer Polytechnic Institute, CII 5015, 110 8th Street, Troy, NY 12180-3590, USA.

APPENDIX B: Ethics in Modeling Workshop Agenda

October 12–14, 1989

Sponsored by the Sloan Foundation and the Department of Decision Sciences and Engineering Systems Rensselaer Polytechnic Institute, Troy, NY, USA

Thursday, October 12

7:00 p.m.   Reception — Inn at the Century, Latham, NY

7:30 p.m.   Dinner — Inn at the Century

8:30 p.m.   Vincent Barabba, General Motors:
                 "The Role of Models in Managerial Decision Making"

Friday, October 13

6:30–8:00 a.m.   Complimentary Breakfast — Inn at the Century

8:30 a.m.   Shuttle to RPI Campus

9:00 a.m.   John Little, Massachusetts Institute of Technology
                 "The Process of Modeling"

                 Discussant: Jay Nunamaker, University of Arizona

10:30 a.m.   Break

10:45 a.m.   Stephen Fienberg, Carnegie-Mellon University:
                 "Statistical Modeling in the Context of the Decennial Census"

                 Discussant: Jim Moor, Dartmouth College

12:15 p.m.   Luncheon

1:15 p.m.   Phil Ross, Environmental Protection Agency:
                  "Ethics, Modeling, and Regulatory Decision Making"

2:00 p.m.   Paul McNelis, Georgetown University:
                  "Rhetoric and Rigor in Macroeconomic Models: Populist and
                  Orthodox Swings in Latin America"

3:30 p.m.   Break

3:34 p.m.   J. Allison, A. Charnes, and W.W. Cooper, University of   Texas
                  "Modeling in Science and Society"

                  Discussant: Thomas Schwartz, University of California at Los
                  Angeles

6:00 p.m.   Reception — Russell Sage Dining Hall, RPI, Troy, NY

6:30 p.m.   Dinner — Russell Sage Dining Hall

7:30 p.m.   Jonathan Rosenhead, London School of Economics and
                  Political Science:
                  "One-Sided Practice — Can We Do Better?"

Saturday, October 14

7:45 a.m.   Shuttle to RPI Campus

8:00 a.m.   Breakfast — RPI

9:00 a.m.   John Mulvey, Princeton University:
                  "Models in the Public Sector: Success, Failure, and Ethical
                  Behavior"

                  Discussant: John Staudenmaier, University of Detroit

10:30 a.m.   Break

10:45 a.m.   Richard Mason, Southern Methodist University:
                  "Morality and Models"

12:15 a.m.   Luncheon — Russell Sage Dining Hall

William A. Wallace, Rensselaer Polytechnic Institute
"Closing Remarks"

2:00 p.m.   Workshop Closing

APPENDIX C: Ethics in Modeling Workshop Participants (Affiliation at Time of Workshop)

PRESENTORS

John Allison, Professor and Director, Center for Legal and Regulatory Studies, Department of Management Science and Information Systems, Graduate School of Business, The University of Texas at Austin, Austin, TX 78712-1175, USA.

Vincent P. Barabba, Executive Director, Department of Marketing and Planning, General Motors Corporation, 3044 West Grand Boulevard, Room 6-235, Detroit, MI 48202, USA.

Abraham Charnes, Professor and Director, Center for Cybernetic Studies, Department of Management Science and Information Systems, Center for Cybernetic Studies, The University of Texas at Austin, CBA 5.202, Austin, TX 78712-1177, USA.

William W. Cooper, Professor of Management, Accounting, Management Science and Information Systems/Nadya Kozemetsky Scott Centennial Fellow, Graduate School of Business, The University of Texas at Austin, Austin, TX 78712-1170, USA.

Stephen E. Fienberg, Maurice Falk Professor of Statistics and Social Sciences and Dean of the College of Humanities and Social Sciences, Carnegie Mellon University, Pittsburgh, PA 15213, USA.

John D.C. Little, Maverick Bunker Professor, Institute Professor, Sloan School, Operations Research Center, Massachusetts Institute of Technology, Cambridge, MA 02319, USA.

Richard O. Mason, Carr P. Collins Distinguished Professor, Department of

Management Information Systems, Cox School of Business, Southern Methodist University, Dallas, TX 75275, USA.

Paul D. McNelis, SJ, Associate Professor, Department of Economics, Georgetown University, Washington, DC 20057, USA.

John M. Mulvey, Director and Professor of Engineering Management Systems, Department of Civil Engineering and Operations Research, Princeton University, Princeton, NJ 08655, USA.

Jonathan Rosenhead, Professor, Operations Research Department, London School of Economics, Houghton Street, London WC2A 2AE, UK.

N. Phillip Ross, Chief of Statistical Policy Branch, Office of Standard Regulations, Department of Policy Planning and Evaluation, United States Environmental Protection Agency, 401 N. Street South West, Washington, DC 20460, USA.

DISCUSSANTS

Deborah Johnson, Associate Professor, Science and Technology Studies, Director of Freshman Studies, 5403 Russell Sage Lab, Rensselaer Polytechnic Institute, 110 8th Street, Troy, NY 12180-3590, USA.

James Moor, Professor, Department of Philosophy, Dartmouth College, Hanover, NH 03755, USA.

Jay Nunamaker, Professor of Management Information Systems and Computer Science, Chairman, Department of Management Information Systems, College of Business and Public Administration, University of Arizona, Tucson, AZ 85721, USA.

Thomas Schwartz, Professor, Department of Political Science, Bunch E Hall, University of California at Los Angeles, Los Angeles, CA 90024, USA.

John Staudenmaier, SJ, Professor, History Department, University of Detroit, 4001 W. McNichols Road, Detroit, MI 48221, USA.

Warren E. Walker, Senior Policy Analyst, Systems Sciences Department, The RAND Corporation, 1700 Main Street, Santa Monica, CA 90406-2138, USA.

PARTICIPANTS

Saul I. Gass, Professor of Business Management, Department of Management Science and Statistics, University of Maryland, Tydings Hall, Room 0141B, College Park, MD 20472-7215, USA.

Ronald Howard, Professor, Department of Engineering and Economic Systems, Terman Engineering Center, Stanford University, Stanford, CA 94305-4025, USA.

Robert O'Keefe, Associate Professor, Department of Decision Sciences and Engineering Systems, Rensselaer Polytechnic Institute, CII 5015, Troy, NY 12180-3590, USA.

Rosalind Pinkus, Associate Professor of Neurological Surgery, History and Ethics, Associate Director for Continuing Education for the Center for Medical Ethics, 9402 Presbyterian University Hospital, University of Pittsburgh, Pittsburgh, PA 15213, USA.

Larry Shuman, Associate Dean of Academic Affairs, School of Engineering, Professor of Industrial Engineering, 323 Bendum Hall, University of Pittsburgh, Pittsburgh, PA 15261, USA.

James M. Tien, Chairman, Department of Decision Sciences and Engineering Systems, Rensselaer Polytechnic Institute, CII 5015, Troy, NY 12180-3590, USA.

Thomas Willemain, Associate Professor, Department of Decision Sciences and Engineering Systems, CII 5207, Rensselaer Polytechnic Institute, 110 8th Street, Troy, NY 12180-3590, USA.

Ann G. T. Young, Business Research Associate, Department of Market Research Technology and Corporate Planning, Eastman Kodak Company, 343 State Street, Rochester, NY 15650-0917, USA.

CONVENOR

William A. Wallace, Professor, Department of Decision Sciences and Engineering Systems, Rensselaer Polytechnic Institute, CII 5015, Troy, NY 12180-3590, USA.

AUTHOR INDEX

Ackoff, R. L. 172, 181, 189, 199, 202, 206
ACM 224
Alexander, E. R. 232, 240
Allen, P. 233, 240
Allison, J. 3, 11
Alter, S. 54
Anderson, J. C. 33
Anderson, M. J. 106, 136
Anderson, R. G. 30, 32, 35
Arnoff, E. L. 172, 181

Bailar, B. A. 126, 139
Banker, R. D. 26, 34
Barabba, V. P. 6, 55, 129–130, 138, 145, 157, 160
Barnett, A. 223
Barnett, V. 44, 56
Baumol, W. J. 15, 31
Bazerman, M. H. 176–177, 182
Bennett, B. 233, 240
Berg, A. 87, 101
Berger, H. 223
Bernstein, R. J. 57
Betz, F. 57
Bishop, Y. M. M. 137–138
Blanning, R. W. 54
Blitzer, H. L. 145, 159
Blount-White, S. E. 37, 54, 73
Boardman, A. E. 27–28, 34–35
Bodily, S. E. 54
Boggs, P. T. 35
Boulding, K. 189, 194

Bounpane, P. 126
Bourguinon, F. 87, 101
Box, G. E. P. 44, 56
Branson, W. H. 87, 101
Brewer, G. D. 225
Brieman, L. 126
Brill, E. D. 225
Brockett, P. L. 33
Brown, R. H. 55–56

Cain, B. 126
Campbell, E. Q. 34
Carrier, H. D. 4, 37
Carrillo, M. 233, 240
Carroll, V. P. 36
Castaneda, H. 54
Caywood, T. E. 223
Chafee, S. B. 224
Chalk, R. 224
Charnes, A. 11, 18–20, 30–36
Checkland, P. 182
Choi, C. Y. 137
Christensen, D. C. 31
Christensen, L. R. 15, 31–32
Chung, D. B. 73
Churchill, N. C. 25, 34
Churchman, C. W. 37–38, 40, 43–45, 54–57, 154, 157, 159–160, 172, 181, 184, 192, 200, 206
Citro, C. F. 136, 139
Clarke, D. 35
Coale, A. J. 139
Cohen, M. L. 136, 139

Coleman, J. 27–28, 34–35
Collins, H. M. 48, 57
Community OR Unit 206
Conk, M. A. 120, 139
Conrad, R. F. 26, 34–35
Cook, S. 201–202, 206
Cooper, W. W. 11, 18–20, 25, 30–36
Cowan, C. D. 116, 137, 139
Crenson, M. A. 145, 159, 225, 233,
    240
Crissey, B. L. 145, 159, 225, 233,
    240
Culliton, B. J. 220, 224
Cummings, D. 31

Dancy, J. 55
Darroch, J. N. 119, 138
Das Gupta, P. 137
Davis, O. A. 27, 34–35, 39
Davis, G. B. 55
deMelo, J. 87, 101
Deming, W. E. 210–211, 223
Dennett, D. C. 54
Dertouzos, J. N. 34
Descartes, R. 40–41, 55
Dewald, W. G. 30, 32, 35
Dewey, J. 11
Diffendal, G. J. 138
Dornbusch, R. 77–78, 80, 83,
    99–100
Drake, A. W. 225
Drake, P. 77, 99
Dror, Y. 214, 224
Duffuaa, S. O. 33
Dunn, W. N. 44, 56–57

Edwards, S. 77–78, 80, 99–100
Eiger, A. 73
Engel, J. 223
Ericksen, E. P. 119–120, 126,
    137–139, 144

Erickson, J. R. 33
Estrada, L. 126
Evans, D. S. 13, 15–17, 20, 23,
    31–33

Fay, R. 118, 126, 137–138
Fellowship for OR 224
Ferguson, R. O. 32
Festinger, L. 47, 57
Field, R. C. 73
Fienberg, S. E. 5, 103, 106, 112,
    126, 128,
    136–140, 143
Finch, D. 101
Fisher, F. M. 126, 133, 139
Fisher, G. 225
Fisher, R. A. 33, 44
Forester, J. 240
Frankel, M. S. 224
Freedman, D. 105, 110, 119–120,
    126, 136–139
Freund, C. 76, 99
Frohman, A. L. 172, 181

Gass, S. I. 7–8, 207, 223, 225
Giere, R. N. 56
Ginzberg, M. J. 38, 54
Glonek, G. F. V. 138
Goeller, B. F. 225, 233, 240–241
Golany, B. 33, 36
Govindarajan, V. 25, 34
Graettinger, J. S. 73
Greenberger, M. 145, 153, 159, 225,
    233, 240

Hale, P. P. 33
Halek, R. 33, 36
Hamilton, D. P. 224
Harris, S. 6, 161
Hayes, P. J. 54

Heckman, J. J. 13, 15–17, 20, 23, 31–33
Helmer, O. 40, 47, 55, 57
Hobson, C. J. 34
Hodges, J. S. 232–233, 240
Hofstadter, D. R. 56
Hogan, H. 137–138
Holland, P. W. 137–138
Holling, C. S. 237, 241
Horton, S. 100
House, P. W. 225
Howard, R. A. 175, 181
Huang, E. T. 138
Hurley, M. W. 54
Husserl, E. 41, 55

Ijiri, Y. 30
Ingersoll, Jr. J. E. 73
Isaki, C. T. 138

Jackson, H. F. 35
Jacobs, J. M. 73
Janis, I. L. 47, 57
Jaro, M. A. 118, 138
Jefferson, T. 108, 136
Johnson, P. E. 46, 57
Joscelyn, K. B. 159
Junker, B. W. 138

Kadane, J. B. 119–120, 137–139, 144
Karp, A. 224
Keeney, R. L. 175, 181, 225, 230, 240
Keyfitz, N. 138
Kilmann, R. H. 44, 47, 55–57
King, J. L. 225
Klopp, G. 36
Kolb, D. A. 172, 181
Kraemer, K. L. 225

Kruskal, W. H. 136
Kuhn, T. 56
Kusnic, M. W. 181–182

LaLonde, R. J. 23–24, 33–34
LaPorte, T. R. 54
Learner, D. B. 35
Leet, E. H. 8, 242
Leonard, T. 44, 56
Little, J. D. C. 6, 54, 181–182
Loveman, B. 78, 99
Lucas, J. 56
Luijpen, W. A. 55

Machol, R. E. 30, 223–224
Magee, J. F. 223
Majone, G. 225
Maki, U. 85, 101
Malec, D. 116, 137, 139
Manheim, M. L. 231, 240
Mann, L. 47, 57
Markie, P. J. 222, 224
Mason, R. O. 6–7, 40, 55, 138, 157, 160, 183–184, 194
McCarthy, J. 54
McCloskey, D. N. 83–85, 100–101
McKeachie, W. J. 27, 34–35
McLeod, J. 225
McNelis, P. D. 5, 75, 88, 100–102
McPartland, J. 34
Meier, P. 126, 133–134, 139
Meltzer, B. 54
Michie, D. 54
Miller, L. 225
Miser, H. J. 223, 231–232, 236, 240–241
Mitroff, I. I. 40, 44, 47, 55–57, 138, 154, 157, 160
Mood, A. M. 34, 227, 240
Morales, J. A. 83, 100

Morande, F. 101
Morris, W. T. 39, 47, 55, 57
MORS 224
Morse, P. M. 225
Moult, W. H. 35
Mulvey, J. M. 5, 37, 39, 54–55, 58,
    73, 196, 206, 222, 224–225,
    238, 241
Mulvey, M. H. 138–139
Munevar, G. 56

Narayanan, A. 57
Nash, S. G. 35
Navidi, W. C. 110, 119–120, 136–
    139
Newell, A. 175, 182
Newton, H. A. 108, 137
Nickelsburg, G. 44, 88, 101

Olson, M. H. 39, 55
ORSA 240-241
Owen, D. 181–182

Panzar, J. C. 15, 31
Passel, J. S. 137
Pepper, S. C. 42–43, 55
Pettigrew, T. F. 228, 230, 233
Phillips, F. Y. 35
Pinkus, R. 245
Poincaré, H. 177, 182
Polanyi, M. 47–48, 57
Pollock, S. 223
Pounds, W. F. 175, 182
Powell, S. 35
Prewitt, K. 136
Pritchett, H. S. 108, 137
Putnam, H. 54

Quade, E. S. 159, 214, 224–225,
    231–232, 236, 240–241

Quetelet, A. 106

Raiffa, H. 53–55, 57, 175, 181
Rappaport, S. 84, 101
Reitman, W. 54
Relman, A. 220
Rescher, N. 40, 47, 55, 57
Reynolds, R. D. 212, 224
Richardson, B. C. 153, 159
Roberts, L. 224
Robinson, J. G. 137
Rolph, J. 126
Rosenberg, A. 84, 101
Rosenhead, J. 7, 195, 206
Ross, N. P. 6, 161
Rothenberg, H. 56
Rothkopf, M. 223
Runes, D. D. 57

Saalberg, J. H. 159
Sachs, J. 77, 87, 95, 99, 101–102
Sagasti, F. 40, 55
Sanday, P. R. 27, 34–35
Sandberg, A. 206
Savage, L. J. 54, 56
Schenker, N. 118, 138
Schmidt-Hebbel, K. 101
Schmitz, E. 36
Schoech, P. E. 31
Schultz, L. K. 138
Seiford, L. 36
Selsor, J. L. 73
Sen, A. K. 101
Shapiro, S. 116, 137
Shaw, J. C. 175, 182
Simon, H. A. 39, 55, 175, 182
Sisson, R. L. 225
Skinner, T. J. 137
Slingerland, C. L. 33
Smith, G. F. 175, 182
Spencer, B. D. 138–139

Sprague, R. H. 54
Stanley, C. M. 216, 224
Staudenmaier, J. M. 9
Steel, D. G. 137
Stigler, S. M. 105, 136–137
Stohr, E. A. 38, 54
Straus, R. P. 26, 34–35
Stutz, J. 36
Sueyoshi, T. 11, 18–20, 31–33, 36
Sutherland, J. W. 46, 56
Swarts, J. 34
Szanton, P. 197, 201, 206

Thomas, D. 33–36
Thrall, R. M. 223
Thursby, J. G. 30, 32, 35
Tracy, W. R. 116, 137
Tufte, E. R. 234, 241
Tukey, J. W. 119, 126, 137–138
Turing, A. M. 45, 56
Turoff, M. 55

Urban, G. L. 172, 174, 181

Vandertulp, J. J. 33
Vidal, R. V. V. 206
Von Bertalanffy, L. 55–56

Wachter, K. 126
Walker, F. A. 137
Walker, W. F. 7–8, 225–226, 231, 233, 240
Wallace, W. A. 1, 8, 37, 54, 242
Wallman, K. K. 139
Warwick, D. P. 228, 230, 233, 240
Waterman, D. A. 46, 56–57
Weinfeld, F. D. 34
Wildavsky, A. 229, 240
Willemain, T. R. 9
Willig, R. D. 15, 31
Winner, L. 39, 54
Wolf, C. 225
Wolfgang, G. S. 138
Wolter, K. M. 105, 116, 119, 126, 136–139, 143
Wu, C. 56
Wyscarver, R. A. 73

Yazdani, M. 57
York, R. L. 34
Young, D. 54

Zachary, W. 54
Zalavsky, A. M. 138
Zellner, A. 32
Zenios, S. A. 73

SUBJECT INDEX

Artificial intelligence (AI) 37–38, 46
    ethical problems inherent in 38

Barabba's Law 6, 145, 152, 159, 244
Brand loyalty 66

Cancer incidences, data relating to
    163–164
Caveat emptor approach for model
    users 152–153, 159
Community operational research
    202–206
Computerized decision
    procedures/(CDPs) 60–70,
    72, 238–239
Constrained regression approach 16,
    18, 20

Decision aid technology 37–57
    decision support systems,
        diagnostic procedure for
        51–53
    epistemological typology of
        decision-aid technologies
        49–51
    epistemology as method of
        inquiry 40–41
    expert systems 37–38, 45–51
    metaphor as a tool 41–44
    statistics and operations research
        44–45

Decision analysis 168, 175–176
Decision making, managerial
        145–160
Decision support systems 51–53

Econometric methods/models
        15–16, 20, 23, 26, 176, 233
Environmental impact assessment
        (EIA) 231, 244
Environmental protection 162–164,
        231, 244
Epistemology as method of inquiry
        40–41
Errors 39, 47, 53, 164–165
    bias 164–165
    systematic error 164
    Types I, II and III 39, 47, 53, 164
Ethical concerns and ethical answers
        207–225
    code of professional conduct,
        need for
        210–212
    codes of ethics/professional
        conduct in related
        professions 210–214,
        216–217
    ethical concerns of students and
        faculty
        219–222
    examples of ethical situations in
        OR modeling 222–223
    modeling process, managing the
        209–210

military OR Society (MORS)
    218
National Institutes of Health
    215–216
Professional Conduct Guide for
    Fellowship of OR
    218–219
Ethics/professional conduct, codes of
    67, 70–73, 147, 152, 159,
    210–219, 243
    for policy scientists 214
    in related professions 210
    from social sciences 212–214
    see Guidelines for Practice of OR
    Guidelines for Professional
        Practice (ORSA)
Expert systems 37–38, 45–51
    epistemological perspective on
    45–49

Goal programming 16, 18, 20–21
Guidelines for the Practice of
    Operations Research
    233–234, 237
Guidelines for Professional Practice
    (ORSA)
    208–209, 215, 218–219, 236,
    238–239

Inquiry Center 154, 157–159
Inquiry systems 154–159
Institute of Management Sciences
    (TIMS) 210
International Monetary Fund model
    79–83, 100

Latin American countries 75–102
    alternative macroeconomics
        framework
        86–102

Catholic Church as redistribution
    institution 96, 99
models, populist and orthodox
    77–86, 95
need for economic analysis and
    policy evaluation 75–77

Managerial decision making, the role
    of models in 145–160
Model builders 6–9, 161–206,
    226–241
    as analysts 227–241
    documentation, need for 165,
        223, 236–237
    ethical/professional/social
        responsibilities of
        6–9, 161–206
Model building 167–182, 232–233
    democratization of 183, 194
    meaning of 167–169
    morality and models 183–194
    for problem solving 170–175
    for science 169–170
    social responsibility in 177–181
    where do models come from?
        175–177
Model users 152–153, 159, 162–166
Modeler as analyst 226–241
    analyzing alternatives and
        drawing conclusions
        234–235
    building models 232–233
    collecting data 233–234
    documenting work and
        communicating results
        236–237
    formulating the problem
        229–230
    identifying alternatives 231–232
    specifying criteria 230–231
Modeling – a one-sided practice
    195–206

'alternative' clientele 199–202
community OR 202–206
differential modeling,
        consequence of 196–197
modeling, motives and managers
        198–200
Modeling, process of 1–3
Models, hazards of 2–3
Models, incorporation of values in
        75–160,
        192–194
Models, some definitions of 13–15
Models in managerial decision
        making 145–160
Models in the public sector 58–73
    CDPs 60–70, 72
    and ethics 59, 65, 67, 70–71
Morality and models 183–194
    covenant with values 192–194
    reality and the model 188–192
    specific projects 184–192

National Institutes of Health
        215–216
    Guidelines for Conduct of
        Research 215

Operations research and management
        science (OR/MS) disciplines
        15–16, 20, 22, 32,
        37–38, 50, 64, 168–172,
        177–178,
        198–202, 222–223
    code of ethics, need for 67
    community OR 202–206
    competing models, evaluation of
        66–67
    conduct of 70–73
    Guidelines, existing 208–209,
        215, 218–219, 233–234,
        236–239

logic and metaphors of statistics
        and operations research
        44–45
military OR 207–208, 218
specific projects 184–188,
        190–192

Philosophic inquiry 40–44, 47–49,
        53
Professional responsibility, evolving
        standards of 161, 165–166,
        177–178, 183–184,
        192–194, 196–198

Regression analysis 39
Research data, accessibility to
        28–30, 215, 220–221
Responsible policy modeling
        226–241
    good practice in each stage of
        analysis
        229–237
    strategy for encouraging
        responsible modeling
        237–241
Risk management 161–166

Science, socially responsible use of
        198
Science, uses of modeling in 11–36,
        169–170
    across-discipline comparison
        15–23
    data and data availability 28–30
    different methods 26–28
    within-discipline methodological
        differences 23–26
Social responsibility in model
        building 177–181
Social sciences, models in 169

Society's role in ethics of modeling
242–245
Statistics/statistical methodologies
15–16, 38,
44–45, 49–50, 105–107
definition of 38
logic and metaphors of statistics
and OR 44–45
models 168

U.S. 1990 census results, ethical and
modeling considerations in
correction of 103–144
1990 U.S. decennial census
103–105
census correction process
143–144
errors found in census
enumeration process
140–142

reflections upon statistical
models/statisticians in
census context 129–136
sampling and statistical models,
early census uses of
142–143
statistical models 105–107
undercount, problem relating to
107–111
dispute on adjustment of
undercount 120–129
statistical techniques for
estimation of 111–120
Users, model
see Model users

Values incorporated in models
75–160, 192–194

Printed in the United Kingdom
by Lightning Source UK Ltd.
105130UKS00002B/49-63